Studies in Systems, Decision and Control

Volume 527

The series "Studies in Systems, Decision and Control" (SSDC) covers both new developments and advances, as well as the state of the art, in the various areas of broadly perceived systems, decision making and control–quickly, up to date and with a high quality. The intent is to cover the theory, applications, and perspectives on the state of the art and future developments relevant to systems, decision making, control, complex processes and related areas, as embedded in the fields of engineering, computer science, physics, economics, social and life sciences, as well as the paradigms and methodologies behind them. The series contains monographs, textbooks, lecture notes and edited volumes in systems, decision making and control spanning the areas of Cyber-Physical Systems, Autonomous Systems, Sensor Networks, Control Systems, Energy Systems, Automotive Systems, Biological Systems, Vehicular Networking and Connected Vehicles, Aerospace Systems, Automation, Manufacturing, Smart Grids, Nonlinear Systems, Power Systems, Robotics, Social Systems, Economic Systems and other. Of particular value to both the contributors and the readership are the short publication timeframe and the world-wide distribution and exposure which enable both a wide and rapid dissemination of research output.

Indexed by SCOPUS, DBLP, WTI Frankfurt eG, zbMATH, SCImago.

All books published in the series are submitted for consideration in Web of Science.

Jamal Mabrouki · Mourade Azrour
Editors

Integrated Solutions for Smart and Sustainable Environmental Conservation

Springer

Editors
Jamal Mabrouki
Mohammed V University
Rabat, Morocco

Mourade Azrour
Moulay Ismail University of Meknes
Errachidia, Morocco

ISSN 2198-4182 ISSN 2198-4190 (electronic)
Studies in Systems, Decision and Control
ISBN 978-3-031-55786-6 ISBN 978-3-031-55787-3 (eBook)
https://doi.org/10.1007/978-3-031-55787-3

This Springer imprint is published by the registered company Springer Nature Switzerland AG
The registered company address is: Gewerbestrasse 11, 6330 Cham, Switzerland

Paper in this product is recyclable.

Contents

Dubins-Based Trajectories: Enhancing Smart Air Traffic Management for Energy-Efficient Environments

Wafae Azoulay, Rafael Casado, Abdelkrim Haqiq, Aurelio Bermúdez, and Luis Orozco

Abstract This study presents an investigation into the use of Dubins trajectories for runway approach procedures and their 4D tracking by aircraft. The study proposes a method to generate a fictitious aircraft that will follow the Dubins trajectory, providing a 4D reference for real aircraft to follow. It explores an innovative approach for 4D trajectory tracking of aircraft in smart environments and energy-efficient settings, based on Dubins-based aircraft approximation trajectories. These trajectories offer optimal curves that connect specified positions while considering the aircraft's turning constraints. The study discusses the theoretical foundations of 4D trajectories and Dubins-based paths, presenting algorithms commonly used in aviation trajectory tracking. A review of existing approaches in the literature is provided, highlighting the limitations and opportunities for improvement. The study then details the proposed approach, emphasizing its potential to optimize aircraft trajectory tracking and reduce fuel consumption and emissions. Simulations and real-world data examples illustrate the efficiency of the approach. Furthermore, the study discusses potential applications in smart environments and energy management, emphasizing benefits such as improved air traffic management and enhanced energy efficiency. It concludes by outlining future research perspectives and challenges,

W. Azoulay (✉) · A. Haqiq
Faculty of Sciences and Techniques, Computer, Networks, Mobility and Modeling Laboratory: IR2M, Hassan First University of Settat, 26000 Settat, Morocco
e-mail: wafae.azoulay@alu.uclm.es; w.azoulay@uhp.ac.ma

A. Haqiq
e-mail: abdelkrim.haqiq@uhp.ac.ma

W. Azoulay · R. Casado · A. Bermúdez · L. Orozco
Albacete Research Institute of Informatics, University of Castilla-La Mancha, 02071 Albacete, Spain
e-mail: rafael.casado@uclm.es

A. Bermúdez
e-mail: aurelio.bermudez@uclm.es

L. Orozco
e-mail: luis.orozco@uclm.es

J. Mabrouki and M. Azrour (eds.), *Integrated Solutions for Smart and Sustainable Environmental Conservation*, Studies in Systems, Decision and Control 527,
https://doi.org/10.1007/978-3-031-55787-3_1

emphasizing the need for integrating the approach into smart air traffic management systems while addressing operational complexities and ensuring data security.

Keywords 4D trajectory tracking · Aircraft precision approach · Navigation error · Dubins path · Smart environments · Energy efficiency · Air traffic management · Aviation

1 Introduction

Modern aviation faces formidable challenges, including the ever-expanding air traffic volume, the escalating demand for energy efficiency, and the imperative to minimize environmental impact. In response, 4D trajectory tracking for aircraft emerges as a groundbreaking and indispensable approach to address these challenges comprehensively. This chapter delves into this advanced approach and its pivotal role within smart environments and energy management [1].

4D trajectory tracking, also known as time–space trajectory tracking, represents a profound evolution in air traffic management. Conventional methods tracked aircraft solely in three-dimensional space, disregarding the temporal aspect of flights. The advent of 4D trajectory tracking allows for the comprehensive description and control of aircraft trajectories in four dimensions: latitude, longitude, altitude, and time. Consequently, the ability to plan, predict, and adapt aircraft trajectories while considering time constraints and environmental conditions is now a reality.

In the context of smart environments, where connectivity, automation, and artificial intelligence play crucial roles, 4D trajectory tracking takes on a new dimension. Technological advancements, such as the Internet of Things (IoT), 5G, machine learning, and autonomous systems, enable real-time communication between aircraft and air traffic control centers. This opens exciting prospects for more precise, proactive, and secure air traffic management, as well as the optimization of flight routes and altitudes [1].

Moreover, energy efficiency has become a top priority for the aviation industry, facing mounting pressures to reduce its environmental footprint. Herein lies the potential of 4D trajectory tracking to significantly contribute to a more sustainable aviation sector. By optimizing aircraft trajectories based on weather conditions, favorable wind patterns, and operational requirements, the approach promises reduced fuel consumption, greenhouse gas emissions, and operational costs, while upholding a paramount level of safety [2].

The core objective of this chapter is to present an innovative approach for 4D trajectory tracking of aircraft, with a particular focus on leveraging Dubins trajectories. Dubins trajectories represent optimal curves that connect two specified positions in three-dimensional space while accounting for the aircraft's limited turning capabilities. This approach offers a promising solution to enhance the precision, reliability, and efficiency of 4D trajectory tracking in complex and dynamic environments.

Throughout this chapter, we explore in detail the underlying theory and principles of 4D aircraft trajectory tracking within smart environments and energy management. We emphasize the advantages of this pioneering approach, showcasing its potential to optimize air traffic, enhance energy efficiency, and foster sustainable aviation practices. Concrete examples and case studies are provided to illustrate the practicality and effectiveness of this compelling approach.

2 Theoretical Foundations

The 4D trajectory tracking for aircraft in smart environments and energy-efficient settings is built upon robust theoretical foundations that combine spatial and temporal aspects of air navigation. To comprehend this innovative approach thoroughly, it is crucial to delve into key concepts of 4D trajectory and Dubins-based trajectories, as well as the algorithms employed in aviation to ensure precise and efficient trajectory tracking.

2.1 The 4D Trajectory and Its Dimensions

The 4D trajectory of an aircraft represents its position in a four-dimensional space: latitude, longitude, altitude, and time. Unlike traditional 3D trajectories, which focus solely on spatial movements, the 4D trajectory takes into account the temporal evolution of the flight. This temporal dimension is crucial for optimizing routes, predicting future trajectories, and thereby reducing delays and potential conflicts between aircraft. By incorporating the time dimension, 4D trajectory tracking enables more precise and dynamic flight planning, ultimately enhancing the overall efficiency of air traffic [3].

2.2 Dubins-Based Trajectories

Dubins-based trajectories, pioneered by Lester Dubins, are integral to 4D trajectory tracking in aircraft operations. These optimal trajectories offer the shortest path between two specified positions while respecting the aircraft's physical limitations, particularly regarding tight turns. The trajectories comprise segments of constant radius turns, connecting the starting and ending points, which helps avoid excessive turns that could lead to unnecessary fuel consumption or compromise flight safety. By employing Dubins-based trajectories, aircraft can follow optimized routes in terms of both time and fuel consumption [4].

A Dubins curve is a trajectory between two points A and B, associated with two orientations and described by a word. When the aircraft operates in an environment

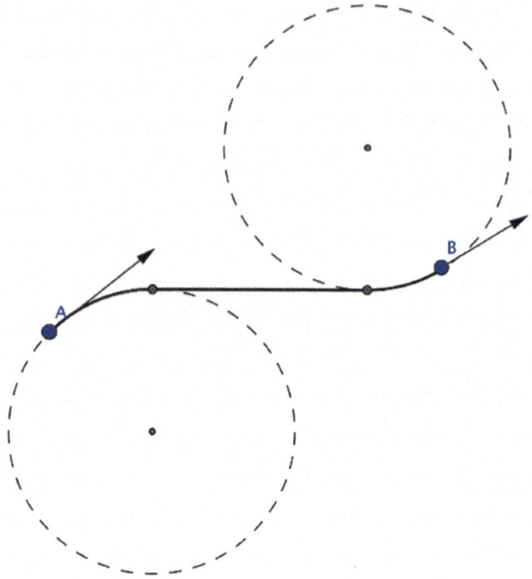

Fig. 1 Example of Dubins curve

with obstacles, each obstacle vertex is considered a subgoal that must be reached. To handle this scenario, the problem is discretized to obtain a set of configurations, with each configuration representing a node in a graph. These nodes are then connected by Dubins curves. Once the graph is constructed, a pathfinding algorithm, such as Dijkstra, is utilized to find a path that navigates through the nodes and the associated Dubins curves [5].

It is essential to note that while the obtained path using the Dijkstra algorithm is free of maneuvers, it may not be globally of minimum length. This implies that the path is efficient in terms of obstacle avoidance and adherence to aircraft constraints but may not represent the shortest path possible. Nonetheless, Dubins-based trajectories remain a valuable tool in optimizing aircraft routes and enhancing operational efficiency in various flight scenarios (Fig. 1).

2.3 *Trajectory Tracking Algorithms*

To efficiently implement and optimize 4D trajectories in aviation, a series of sophisticated algorithms have been developed. These algorithms leverage mathematical optimization methods, adaptive control, and advanced trajectory planning techniques. Some of these algorithms are integrated into Air Traffic Management (ATM) systems and air navigation systems for real-time use. They consider airspace constraints, flight regulations, as well as operational preferences of airlines. These algorithms enable effective air traffic management, minimizing potential conflicts, and optimizing trajectories for enhanced energy efficiency and reduced carbon emissions.

The theoretical foundations of 4D trajectory tracking for aircraft rely on a four-dimensional representation of aircraft positions, integrating the temporal dimension for precise and dynamic route planning. Dubins-based trajectories provide an optimal approach to connect specified positions while avoiding tight turns. The advanced algorithms used in aviation allow for the practical implementation of these theoretical concepts, providing advanced solutions to improve air traffic management in smart and energy-efficient environments.

3 Review of Existing Approaches

Researchers and aviation experts have made significant progress in developing methods to implement and optimize 4D trajectories, aiming to enhance air traffic management and improve energy efficiency.

Numerous studies have explored different techniques and algorithms for 4D trajectory tracking [6, 7]. While traditional approaches primarily focused on 3D navigation with limited consideration of time, recent advancements have incorporated the fourth dimension of time to improve trajectory prediction and planning. These approaches range from basic interpolation methods to more sophisticated machine learning algorithms, considering weather conditions, aircraft performance, and traffic flow. Some studies have also investigated the use of probabilistic models to address prediction uncertainties, thereby enhancing overall safety and adaptability in dynamic airspace environments. Additionally, integrating ground-based and satellite-based navigation systems has further bolstered the accuracy and reliability of 4D trajectory tracking.

3.1 *Identification of Limitations and Improvement Opportunities*

While there have been significant advancements, the current approaches still encounter certain limitations. One major challenge is the intricate integration of 4D trajectory tracking into existing air traffic management systems, ensuring seamless coordination among multiple aircraft and air traffic control centers. Additionally, the real-time nature of air traffic requires rapid and efficient trajectory adjustments, calling for further developments in predictive algorithms and real-time data processing capabilities.

Furthermore, ensuring the confidentiality and security of data remains a concern, as precise aircraft trajectory information could be potentially misused or pose security risks. Additionally, the variability of international regulations and airspace management practices represents a barrier to the widespread implementation of 4D trajectory tracking on a global scale.

However, these limitations also present opportunities for improvement. Future research can focus on enhancing the robustness and adaptability of 4D trajectory tracking algorithms, exploring advanced machine learning models and artificial intelligence techniques to improve trajectory prediction accuracy. Moreover, the integration of advanced communication systems and IoT technologies can facilitate real-time data sharing between aircraft and air traffic control centers, enabling smoother trajectory adjustments and conflict resolution.

4 Innovative Approach: Dubins-Based Aircraft Approximation Trajectories

The innovative approach proposed in this chapter is based on the use of Dubins-based trajectories to significantly enhance aircraft trajectory tracking in intelligent and energy-efficient environments. Dubins trajectories are optimal curves that connect two specified positions while adhering to the aircraft's limited turning constraints. This approach offers a promising solution to improve the accuracy, reliability, and efficiency of 4D trajectory tracking, while simultaneously reducing fuel consumption and greenhouse gas emissions.

The method involves generating approximation trajectories based on Dubins curves for each aircraft, considering operational constraints such as maximum speed, minimum altitude, and turning limitations. Advanced trajectory planning techniques are employed to optimize Dubins trajectories, minimizing flight time and distance while avoiding excessively tight turns or abrupt altitude changes. Furthermore, the approach considers weather conditions and favorable winds to adjust trajectories and achieve high levels of energy efficiency.

To illustrate the effectiveness of the proposed approach, detailed simulations are conducted using advanced flight models and real data from commercial flights. Simulation results demonstrate that Dubins-based trajectories lead to a significant reduction in fuel consumption compared to conventional trajectory approaches. Aircraft can follow smoother and optimized trajectories, avoiding unnecessary altitude fluctuations and minimizing flight times. Moreover, the simulations show that the use of Dubins trajectories improves flight predictability, reducing delays and air traffic congestion.

By utilizing concrete examples based on real data, this chapter highlights the efficiency and potential of the innovative approach using Dubins-based aircraft approximation trajectories. These encouraging results strengthen the idea that this approach can play a crucial role in air traffic management in intelligent environments, allowing precise optimization of aircraft trajectories while contributing to reducing the aviation industry's environmental footprint. By integrating Dubins-based trajectories into air traffic management systems and adopting them on a large scale, the aviation industry can aspire to more efficient, safe, and sustainable flight management in future environments.

4.1 Potential Applications in Intelligent and Energy-Efficient Environments

The innovative approach of Dubins-based aircraft approximation trajectories offers a range of practical applications in aviation and intelligent transportation. In intelligent environments, where connectivity, automation, and artificial intelligence play a crucial role, adopting Dubins trajectories has the potential to revolutionize air traffic management and enhance the energy efficiency of the aviation sector.

Regarding air traffic management, the adoption of Dubins trajectories would significantly reduce delays and traffic congestion by optimizing smooth and direct routes. Aircraft could follow more efficient paths, avoiding unnecessary detours and excessive turns. This would lead to an overall reduction in flight time and operational costs for airlines while improving the passenger experience.

In terms of energy efficiency, using Dubins trajectories would enable better fuel consumption management for aircraft. By optimizing routes based on weather conditions, favorable winds, and operational constraints, aircraft could save fuel and reduce CO_2 emissions. This contributes to the aviation industry's overall goal of becoming more sustainable and environmentally friendly.

Moreover, integrating Dubins trajectories into air traffic management systems would facilitate better flight coordination and reduce potential conflicts between aircraft. Advanced real-time algorithms could adjust trajectories based on changing weather conditions, unforeseen delays, or airspace constraints. This would enhance flight safety and strengthen the resilience of the air traffic management system in complex and dynamic situations.

In summary, the innovative approach of Dubins-based aircraft approximation trajectories offers a promising range of applications in intelligent and energy-efficient environments. By optimizing routes, improving energy efficiency, and enhancing air traffic management, this approach can contribute to making aviation more efficient, sustainable, and intelligent. These potential benefits underscore the importance of continuing research and development in this field to fully realize the advantages of this innovative approach in the future of aviation and intelligent transportation.

5 Methodology

Dubins trajectories are mathematical curves that describe the movement of an aircraft on the final approach [8]. By using these curves, it is possible to define precise trajectories that aircraft must follow to land safely. This reduces the distances aircraft travel in the event of a missed approach, as Dubins' trajectories can be easily readjusted to allow for a new landing attempt more quickly. By using Dubins trajectories, aircraft can also follow more accurate and shorter trajectories, reducing the potential for tracking errors and reducing unnecessary flight distances. In summary, the use of

Dubins trajectories can help improve the efficiency and safety of aircraft approach and landing operations.

A method is proposed to define a fictitious aircraft that will describe Dubins trajectories generating a 4D reference and two tracking mechanisms (basic) following this reference, which can be used by real (flyable) aircraft. These mechanisms have been evaluated in the Málaga airport (Spain), which has a high traffic density during the seasonal period, simulation results with MATLAB software verify their correct operation and high tracking.

6 Precision Approaches

Approach [9] is a critical flight operation with potentially fatal consequences if mishandled. There are two main types of approach procedures: visual and instrument approaches. In visual approaches, the pilot navigates to the airport using visual references. In contrast, instrument approaches involve following a published Instrument Approach Procedure (IAP), which is detailed in an IAP chart or approach plate (see Fig. 2).

The IAP consists of several segments: the initial, intermediate, and final approach segments, and the missed approach segment. Each segment is defined by designated

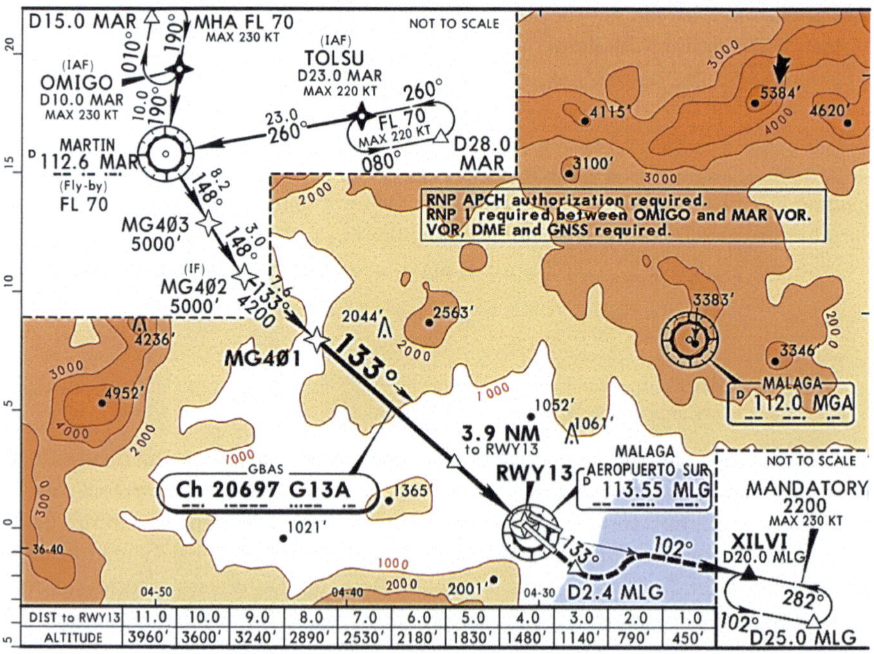

Fig. 2 Detail of the IAP chart for the Málaga "RWY 13" runway [8]

fixes, which are specific geographical positions. During the initial segment, starting at the Initial Approach Fix (IAF), the aircraft transitions from an en-route airway to the intermediate segment. The next segment, starting at the Intermediate Fix (IF), allows the aircraft to descend to an intermediate altitude and align with the runway.

In the final approach segment, commencing at the Final Approach Point (FAP), the aircraft navigates towards the runway using navigation aids situated at or near the runway. When the aircraft reaches the Missed Approach Point (MAPt) or the Decision Altitude/Height (DA/H)—depending on whether it's a no precision or precision approach—it must either proceed with the landing or initiate the missed approach maneuver.

During a non-precision final approach, the aircraft relies on lateral (2D) guidance, typically provided by a VOR station. On the other hand, a precision final approach involves both lateral and vertical (3D) guidance, commonly provided by an Instrument Landing System (ILS) [10]. The ILS incorporates a localizer for lateral guidance and a glideslope for vertical guidance.

7 Evaluation

The proposed methodology aims to improve the efficiency and safety of aircraft approach and landing operations by using Dubins trajectories to define precise and safe landing trajectories. To evaluate the effectiveness of the methodology, a simulation was carried out using MATLAB software, which involved the use of a fictitious aircraft generating a 4D reference trajectory using Dubins curves, along with two tracking mechanisms (basic) that follow this reference trajectory. The evaluation was conducted at the Málaga airport in Spain, which experiences high traffic density during the seasonal period. The simulation results demonstrated that the tracking mechanisms operate correctly and achieve high tracking accuracy when following the Dubins trajectories generated by the fictitious aircraft. This suggests that the proposed methodology could be effective in improving the efficiency and safety of aircraft approach and landing operations in real-world scenarios. By reducing the distances travelled by aircraft during missed approaches and allowing for more accurate and shorter trajectories, the use of Dubins trajectories could potentially reduce the risk of accidents and increase the capacity of busy airports like Málaga.

8 Results and Discussion

We have created a simulation tool using Matlab/Simulink R2023b [11]. The purpose of this tool is to accurately model the air traffic flow approaching a specific airport runway, incorporating all aspects of the approach and landing maneuver. To achieve this, we have utilized a combination of time-based continuous and discrete-event simulation resources available in Simulink.

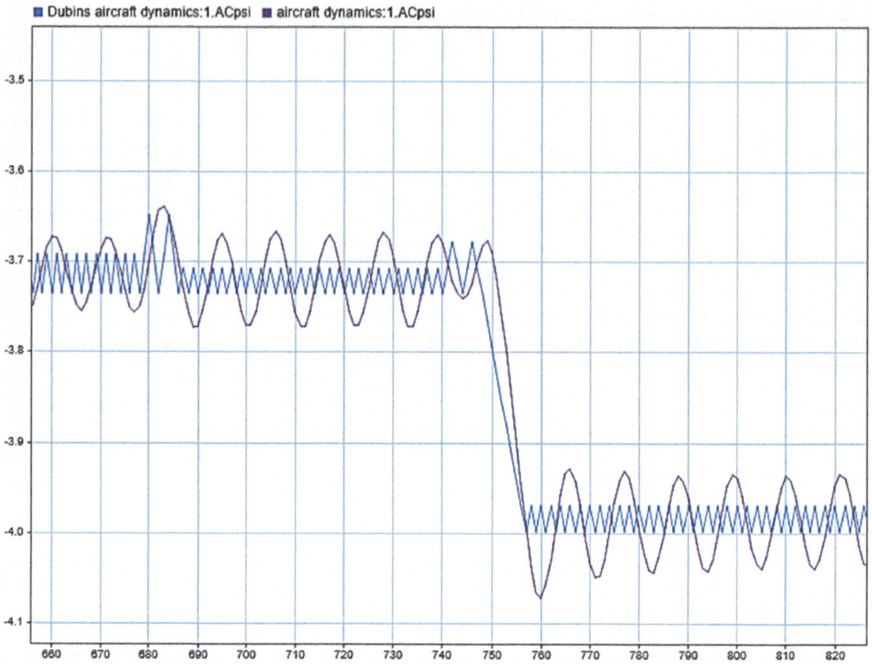

Fig. 3 Flyable aircraft following its Dubins reference

Figure 3 shows a detail of how the aircraft's heading varies during the maneuver. The horizontal axis represents the time of flight, expressed in seconds. The vertical axis represents the aircraft heading, expressed in cumulative radians of difference from the north. A heading less than –pi radians indicates that we have turned left beyond the south direction. The blue series represents the heading of the reference Dubins aircraft. We can observe a small turn of 15° to the left, executed in 10 s; going from heading -3.72 radians (146°) to -3.98 radians (131°). Before and after the turn, the aircraft is moving in a straight line. However, this is not an ideal straight line. The aircraft performs a smooth roll to left and right alternatively, constantly correcting the course while keeping it as straight as possible. The purple series represents how the flyable aircraft tries to follow the Dubins aircraft. The small corrections made by this are magnified in the follow-up obtained.

9 Future Perspectives and Challenges

The 4D trajectory tracking for aircraft in intelligent and energy-efficient environments opens up numerous exciting research opportunities for the future. Due to its potential to enhance air traffic efficiency, and reduce fuel consumption, and carbon emissions, this innovative approach has garnered increasing interest in the aviation and intelligent transportation community. To fully leverage this approach and address upcoming challenges, several research directions emerge.

Integration into Intelligent Air Traffic Management Systems: One of the key future perspectives is the deeper integration of Dubins-based 4D trajectories into intelligent air traffic management systems. This will require efforts to develop advanced planning and control algorithms capable of efficiently managing dense and dynamic air traffic in complex environments.

Multi-Objective Optimization: Future research can focus on multi-objective optimization, considering not only energy efficiency and emission reduction but also other factors like safety, passenger comfort, and operational costs. This holistic approach would enable finding optimal trade-offs between different objectives, thus improving decision-making in air traffic management.

Integration of Emerging Technologies: The rapid evolution of emerging technologies, such as unmanned aerial vehicles (UAVs) and urban air mobility (UAM), presents new perspectives for 4D trajectory tracking. Future research can concentrate on the seamless integration of these technologies into the global air traffic management system to optimize airspace utilization and enhance aerial mobility.

However, despite promising perspectives, there are challenges to address to make this approach more robust and widely applicable:

Operational Complexity: The practical implementation of 4D trajectory tracking on a large scale requires solving complex operational problems, such as traffic management in densely populated areas or changing weather conditions. This demands close collaboration between regulators, airlines, aircraft manufacturers, and air navigation service providers.

Data Security and Privacy: The use of Dubins-based trajectories requires secure management of flight data to preserve the confidentiality of sensitive information while ensuring safe and efficient flight planning. Future research must address these issues to ensure public acceptance and trust.

Costs and Deployment: Widespread adoption of this innovative approach may require significant investments in infrastructure, personnel training, and equipment. Future research should consider these aspects to ensure the approach remains economically viable.

10 Conclusion

This chapter examined the innovative approach of Dubins-based aircraft approximation trajectories to enhance aircraft trajectory tracking in intelligent and energy-efficient environments. The main conclusions drawn from this study emphasize the importance and significant potential of this approach for the future of aircraft trajectory tracking.

In summary, the proposed approach relies on Dubins-based trajectories, which are optimal curves connecting two specified positions while adhering to aircraft turning constraints. This approach offers several significant advantages. It improves the accuracy and efficiency of 4D trajectory tracking, thereby reducing delays, traffic congestion, and carbon emissions. Additionally, it enables optimized fuel consumption management, making aviation more sustainable and environmentally friendly.

The conducted simulations and examples based on real data demonstrated the relevance of this approach, highlighting its capability to reduce aircraft fuel consumption while enhancing flight predictability. These encouraging results reinforce the notion that the Dubins-based aircraft approximation trajectory approach can play a crucial role in future air traffic management, particularly in intelligent and energy-efficient environments.

This innovative approach holds substantial potential for the future of aircraft trajectory tracking. It provides practical solutions to optimize aircraft routes, reduce delays, improve energy efficiency, and enhance flight safety. To make this approach widely applicable, it is essential to address operational challenges, ensure data security, and foster collaboration among aviation stakeholders. With concerted efforts, the future of aircraft trajectory tracking will be characterized by smarter, more sustainable, and more efficient management, benefiting the entire aviation industry and our environment.

References

1. Guan, X., Lv, R., Sun, L., Liu, Y.: A study of 4D trajectory prediction based on machine deep learning. In: Proceedings of the World Congress on Intelligent Control and Automation (WCICA), pp. 24–27, 2016. https://doi.org/10.1109/WCICA.2016.7578458
2. Hammad, A.W.A., David, R., Bu-Qammaz, A., Grzybowska, Hanna, Akbarnezhad, A.: Mathematical optimization in enhancing the sustainability of aircraft trajectory: A review. Int. J. Sustain. Transp. **14**(6), 413–436 (2020)
3. Klooster, J.K., Wichman, K.D., Bleeker, O.F.: 4D trajectory and time-of-arrival control to enable Continuous Descent Arrivals. AIAA Guid. Navig. Control. Conf. Exhib. (2008). https://doi.org/10.2514/6.2008-7402
4. Lin, Y., Saripalli, S.: "Path planning using 3D Dubins Curve for Unmanned Aerial Vehicles. In: 2014 International Conference on Unmanned Aircraft Systems (ICUAS), pp. 296–304, 2014. https://doi.org/10.1109/ICUAS.2014.6842268.
5. Mohanad Al Nuaimi, Analysis and comparison of clothoid and Dubins algorithms for UAV trajectory generation, West Virginia University, 2014
6. Choi, H. J.: Time-optimal paths for a dubins car and dubins airplane with a unidirectional turning constraint. Thesis. University of Michigan, 1–110, 2014
7. Alam, A.S., Hu, Y.-F., Pillai, P., Xu, K., Baddoo, J.: Optimal datalink selection for future aeronautical telecommunication networks. IEEE Trans. Aerosp. Electron. Syst. **53**(5), 2502–2515 (2017)
8. Casado, R.: Manuel L´opez-Lago, Jos´e Serna, and Aurelio Berm´udez, "Enhanced missed approach procedure based on aircraft reinjection." IEEE Trans. Aerosp. Electr. Syst. **57**(6), 4149–4170 (2021)
9. FAA, Instrument Procedures Handbook (IPH), https://www.faa.gov/regulations_policies/handbooks_manuals/aviation/instrument_procedures_handbook/ (2017). Accessed 18 Dec 2020

10. Moir, I., Seabridge, A., Jukes, M.: Civil avionics systems. Wiley (2013)
11. The MathWorks Inc, "Matlab." https://www.mathworks.com/products/matlab.html. Accessed 18 Jan 2023

Improved Salt Stress Tolerance of *Salvia Officinalis* Grown in the Presence of a Halophytic Plant *Spergularia Maritima*: Analysis of Morpho-Physiological Parameters

Issam El-Khadir, Safae Ktaoui, Yassine Mouniane, Ahmed Chriqui, Jamal Mabrouki, Halima Ameziane, and Driss Hmouni

Abstract Our study presents an innovative approach for soil salinity management by utilizing a strategy based on the association of sage (*Salvia officinalis*) with halophytic plants. In this study, we evaluated the effects of salt stress induced by sodium chloride on the morpho-physiological characteristics of sage, both in the presence and absence of the halophyte *Spergularia maritima*. The results obtained demonstrated promising outcomes for the combination of sage with *Spergularia maritima*. The presence of *Spergularia maritima* resulted in a significant increase in root length, particularly at a concentration of 5 g/l, where the root length reached 14.69 cm compared to 11.69 cm for sage alone. A similar trend was observed for the aerial part length, which reached 7.06 cm when combined with *Spergularia maritima*, in contrast to 5.69 cm for sage alone. Moreover, the root volume also exhibited an increase, from 1.36 cm^3 for sage alone to 2.10 cm^3 when combined with *Spergularia maritima*. Additionally, there was an increase in water content in the leaves, stems, and roots for the combined treatment compared to sage alone. The total soluble sugar content was also reduced when sage was planted with *Spergularia maritima*, reaching a value of 0.04 ug/MF compared to 0.06 ug/MF for sage alone at a concentration of 10 g/l. These findings demonstrate that the association of sage with halophytic plants offers a promising solution for

I. El-Khadir (✉) · S. Ktaoui · Y. Mouniane · A. Chriqui · H. Ameziane · D. Hmouni
Laboratory of Natural Resources and Sustainable Development, Faculty of Sciences, Ibn Tofaïl University—KENITRA-University Campus, 14000 Kenitra, Morocco
e-mail: issam.elkhadir@uit.ac.ma

Y. Mouniane
e-mail: yassine.mouniane@uit.ac.ma

J. Mabrouki
Laboratory of Spectroscopy, Molecular Modelling, Materials, Nanomaterials, Water and Environment, CERN2D, Mohammed V University in Rabat, Faculty of Science, AV Ibn Battouta, Agdal, 10106 Rabat, Morocco

J. Mabrouki and M. Azrour (eds.), *Integrated Solutions for Smart and Sustainable Environmental Conservation*, Studies in Systems, Decision and Control 527,
https://doi.org/10.1007/978-3-031-55787-3_2

reducing soil salinity and creating a more favorable environment for crop growth. This innovative approach can contribute to enhancing agricultural productivity in regions facing increasing soil salinity challenges. Moreover, it offers exciting prospects for sustainable management of saline soils, promoting resilient and environmentally friendly agriculture.

Keywords *Salvia officinalis* · *Spergularia maritima* · Sodium chloride · Morpho-physiological characteristics

1 Introduction

Climate change has become one of the most important issues facing the world. One of the most pressing challenges we face is water scarcity, which is exacerbated by increasing salinity concentrations in water resources and soils. This increase in salinity represents a serious threat to agriculture, biodiversity, ecological balance and economic performance [1].

Throughout their lifecycle, higher plants are continuously exposed to numerous biological and abiotic stresses, salt stress being one of the main global climatic factors [2, 3]. Approximately 8.31109 h of the world's land are salinized, which is a significant type of land degradation, and estimates suggest that soil salinity has a significant negative impact on approximately 7.5% of the world's land [4, 5]. Soil salinization is a serious environmental constraint that has an impact on the development and yield of important crops worldwide [6]. In addition, industrial pollution, poor agricultural practices and climate change are leading to a 10% annual increase in soil salinization worldwide [7]. Increased soil salinization affects soil respiration, microbial population stability and soil structure, which indirectly impacts soil nutrient cycling and availability [8].

Plants have developed two different types of defense system against salinity. The first is to control the concentration and distribution of salt within the plant itself, while the second is to limit salt penetration through the roots [9]. In addition, a number of cellular and genetic processes mitigate the effects of salinity on the plant. Increased production of secondary metabolites and altered enzymatic activity are two such processes [10]. To reduce soil salinity and increase plant resistance to salt, efforts have been made to develop a method that is both affordable and environmentally friendly [7–11]. As a result, halophytes have developed a variety of morphological, physiological and biochemical adaptations [12]. Despite a natural tolerance to salt, the level of tolerance of halophytes is strongly regulated by their genetic make-up and stage of development [13].

Salvia officinalis (also known as sage) is a fragrant Mediterranean herb that has spread around the world through naturalization [14]. It has been one of the most popular medicinal plants since Roman times [15]. *Spergularia maritima* a halophilic plant species of salt-influenced primary environments in Central Europe, is currently extending its range to roadside areas that receive de-icing salt treatments in winter [16].

This study explores an assessment of the effectiveness of *Spergularia maritima* in mitigating the negative effects of salt stress on Salvia officinalis by subjecting sage seedlings to high salinity conditions in the presence and absence of the halophytic plant. The underlying idea is that the presence of halophytic plants could bring substantial benefits to salt-sensitive sage by reducing salinity stress and improving its tolerance to salt stress.

2 Material and Methods

2.1 Plant Material and Salt Treatments

Three-month-old rooted cuttings of *Salvia officinalis* and the plant *Spergularia maritima* were grown in plastic pots containing a substrate mix of soil, compost and peat for all treatments (Fig. 1).

The experiment was carried out in a greenhouse at Ibn Tofail University, Morocco, and the halophyte plant *Spergularia maritima* was collected from Oued Malah (Fig. 2).

Two experiments were carried out. In the first experiment, sage was grown alone, and in the second, sage was grown with *Spergularia maritima*. Plants were subjected to different salt treatments using sodium chloride (NaCl) concentrations of 2 g/L (34 mmol/L), 5 g/L (85 mmol/L) and 10 g/L (170 mmol/L).Each group of plants received one of these treatments by adding the corresponding solution to the irrigation water. Watering was carried out in the morning, with a quantity of 100 ml per pot, and the watering frequency varied from 2 to 3 days according to the plants' needs (Fig. 3).

Fig. 1 **a** *Salvia officinalis*; **b** *Spergularia maritima*; **c** *Salvia officinalis* + *Spergularia maritima*

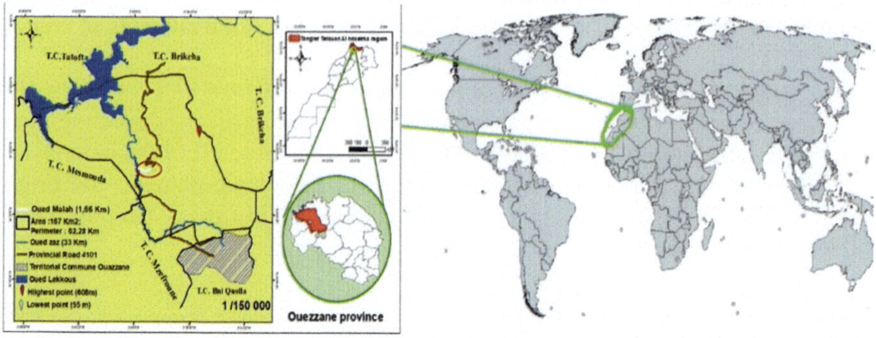

Fig. 2 Location of Oued Malah

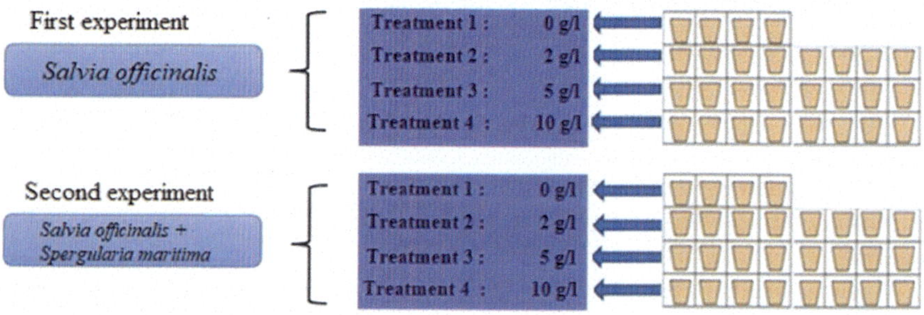

Fig. 3 Illustration of two experiments

2.2 Measured Parameters

Overhead and Underground Lengths

Sage (*Salvia officinalis*) plants were carefully dug up, their roots washed under running water and then quickly wrung out with filter paper. Aboveground (stems + leaves) and root parts were separated. The lengths of aerial and root parts were measured with a ruler.

Root Volume

It is measured by immersion in a graduated test tube using the Archimedes' buoyancy protocol [17].

Number of Leaves

When the first leaflet appears, visual counting of budding leaves begins.

Relative Water Content

Sage plants were harvested, and their fresh weight recorded. After removing excess surface water, the plants were placed in an oven at a constant temperature of 80 °C for a drying period of 24 h, until the dry weight had stabilized. By subtracting the dry weight from the fresh weight, we obtained the amount of water contained in each plant. This difference is expressed as a percentage of fresh matter using the following formula:

$$RWC = \frac{(P_f - P_s)}{P_f} \times 100.$$

With: RWC: Relative Water Content of Plants (%), P_f: Fresh Weight just after harvest (g), P_S: Dry Weight after oven drying (g).

Total Soluble Sugars
Total soluble sugars are determined using the phenol method of Dubois [18]. Absorbance is read in a spectrophotometer at a wavelength of 490 nm, and the soluble sugar content is calculated by reference to a standard glucose curve.

Statistical Analysis
Measured data were subjected to a one-way analysis of variance (ANOVA) using SPSS statistical software. A comparison of means between control and treated samples was then carried out using the post-hoc test. This test can be used to determine significant differences between group means in an analysis of variance. Results are considered significant when $P \leq 0.05$.

3 Results

3.1 *Effect of Salt Stress on the Aerial Length of Sage in the Presence and Absence of* Spergularia Maritima *(Halophyte Plant)*

These results indicate that salt stress has a negative effect on the aerial length of sage, either in the presence or absence of *Spergularia maritima*. However, in the presence of *Spergularia maritima*, sage appears to have a slightly greater aerial length than the control group without salt stress. However, even in the presence of *Spergularia maritima*, increasing NaCl concentration leads to a decrease in the aerial length of sage (Figs. 4 and 5).

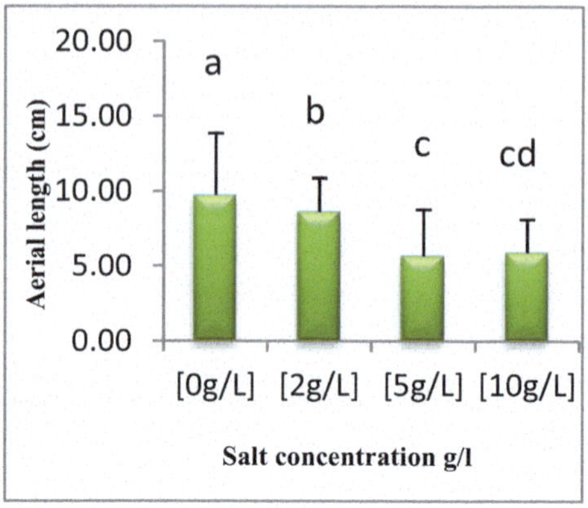

Fig. 4 Variation in the aerial length of sage under different long-term NaCl concentrations

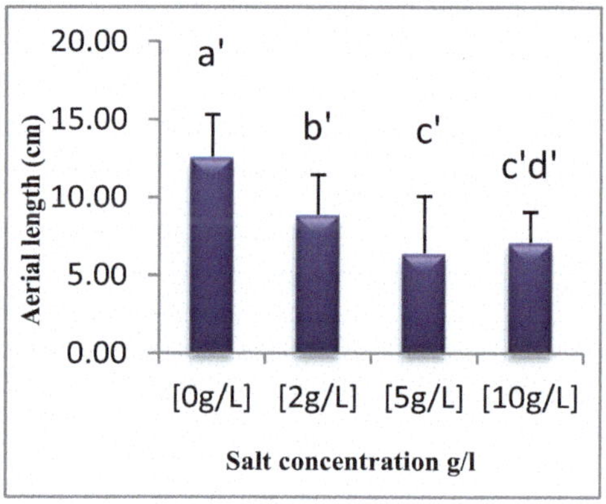

Fig. 5 Variation in the aerial length of Sage under the effect of different NaCl concentrations in the presence of *Spergularia maritima* over the long term

3.2 Effect of Salt Stress on Sage Root Length in the Presence and Absence of Spergularia Maritima *(Halophyte Plant)*

The results show that salinity reduced the length of the root section (LPR) in the plants. The reduction became more pronounced at higher salt concentrations. Salt stress has a negative effect on root growth. Analysis of variance also revealed a significant effect between control and stressed treatments (P: 0.030) (Figs. 6 and 7).

These results indicate that salt stress has a negative effect on sage root length, both in the presence and absence of *Spergularia maritima*. The presence of *Spergularia maritima* appears to have a slight protective effect, as sage root length is slightly greater in the presence of this halophyte plant than in the control group without salt stress. However, even in the presence of *Spergularia maritima*, increasing NaCl concentration led to a decrease in sage root length.

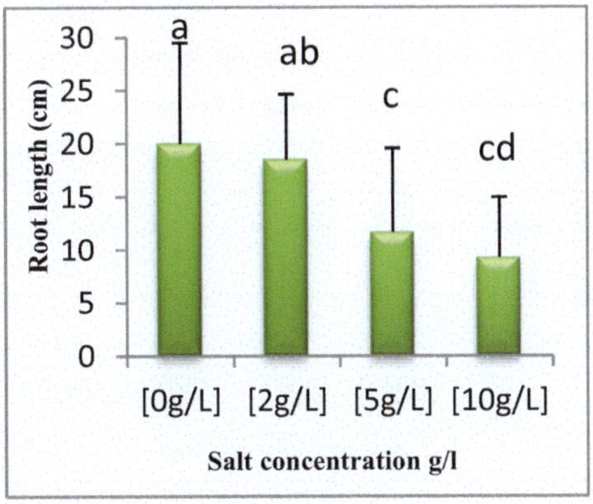

Fig. 6 Variation in root length of sage under different long-term NaCl concentrations. (Values with different letters are significantly different: $p < 0.05$.)

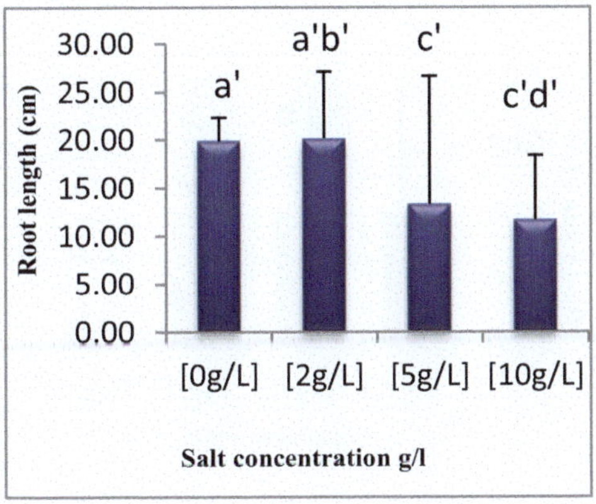

Fig. 7 Variation in root length of Sage under different NaCl concentrations in the presence of halophytic plants (values with different letters are significantly different: $p < 0.05$)

3.3 Effect of Salt Stress on Sage Root Volume in the Presence and Absence of Spergularia Maritima *(Halophyte Plant)*

The results showed that root volume evolved much more in control plants than in those treated with saline solution, which may be attributed to the negative effect of salinity on root branching and the poor adaptation of sage to saline environments. Statistical analysis revealed a highly significant difference between control and stressed treatments (p: 0.003). Statistical analysis of root volume also revealed significant differences between controls and stressed treatments in the presence of *Spergularia maritima* (p: 0.035) (Figs. 8 and 9).

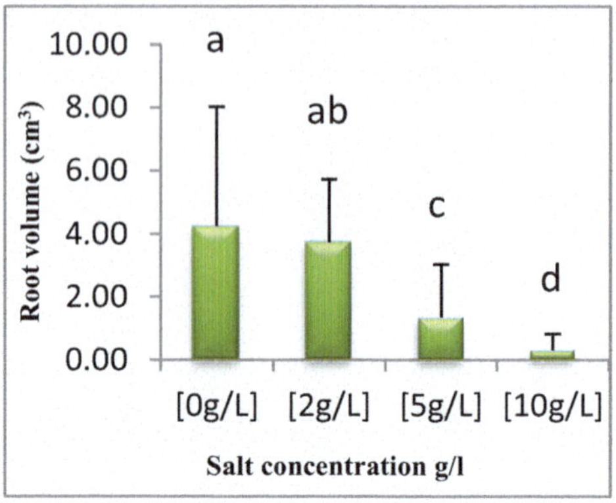

Fig. 8 Change in root volume of sage after 20 days of salt treatment. (Values with different letters are significantly different: $p < 0.05$)

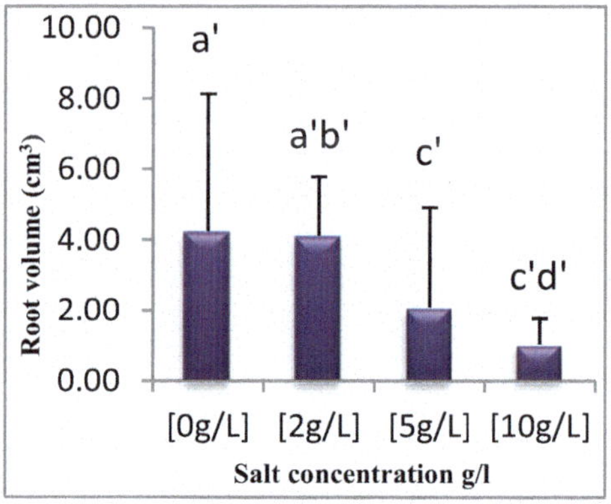

Fig. 9 Variation in root volume of sage under different long-term NaCl concentrations in the presence of halophytic plants (values with different letters are significantly different: $p < 0.05$)

In the presence of *Spergularia maritima*, sage reached the highest root volume of 4.25 cm^3 for the control group, then gradually decreased with increasing salt concentration. At a concentration of 2 g/L NaCl, the average root volume was 3.75 cm^3. At a concentration of 5 g/L NaCl, mean root volume decreased further to 1.36 cm^3. Finally, at a concentration of 10 g/L NaCl, the average root volume reached its lowest level, at 0.33 cm^3.

3.4 *Effect of Salt Stress on the Water Content of Sage Roots in the Presence and Absence of* **Spergularia Maritima** *(Halophyte Plant)*

The results show a significant variation in the water content of sage roots as a function of the different NaCl concentrations used. In the absence of *Spergularia maritima*, the water content of sage roots decreased progressively with increasing NaCl concentrations. It is 87.39% for the control group (0 g/L NaCl), 88.74% at 2 g/L, 76.78% at 5 g/L and 70.67% at 10 g/L (Figs. 10 and 11).

However, in the presence of *Spergularia maritima*, the water content of sage roots is higher in the control group (92.40%) and decreases significantly with increasing NaCl concentrations. It is 88.57% at 2 g/L, 70.31% at 5 g/L and increases slightly to 85.10% at 10 g/L.

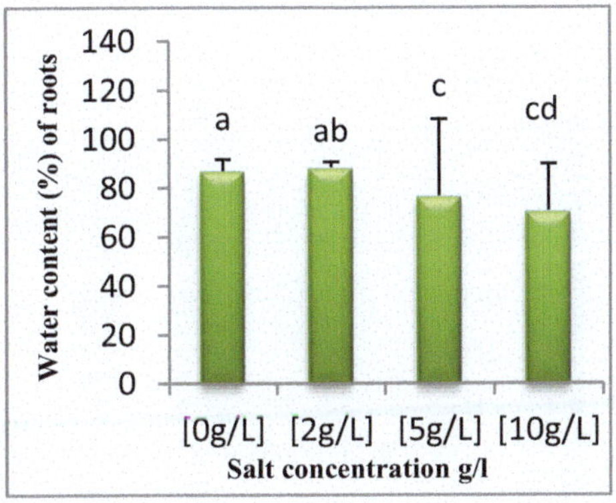

Fig. 10 Variation in water content of sage roots after 20 days of salt treatment (values with different letters are significantly different: $p < 0.05$)

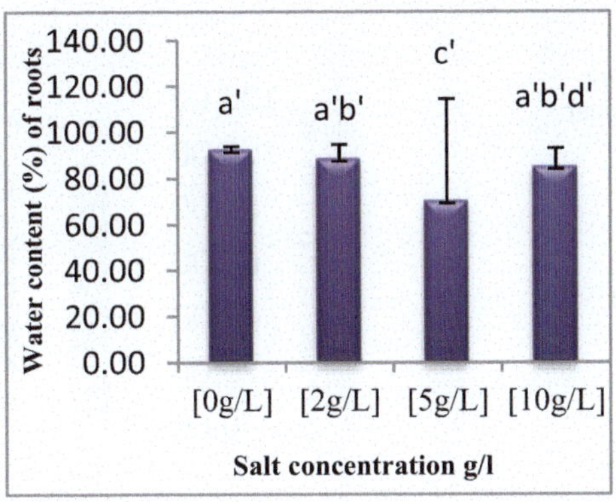

Fig. 11 Variation in water content of sage roots under different NaCl concentrations in the presence of halophytic plants (values with different letters are significantly different: $p < 0.05$)

3.5 *Effect of Salt Stress on the Water Content of Sage Leaves in the Presence and Absence of* Spergularia Maritima *(Halophytic Plant)*

The results show a significant reduction in leaf water content compared with control plants. The higher the NaCl concentration, the lower the leaf water content. Statistical analysis confirmed a significant difference between controls and treatments stressed with 5 g/l and 10 g/l NaCl (P: 0.027), indicating that salt stress has an impact on leaf water content in sage (Figs. 12 and 13).

However, when we examine the results in the presence of *Spergularia maritima*, we note that the water content of sage leaves remains relatively stable for the different NaCl concentrations, with a slight decrease observed for plants irrigated with 10 g/l NaCl. Statistical analysis revealed no significant differences between control and stressed treatments in the presence of *Spergularia maritima* (P: 0.19).

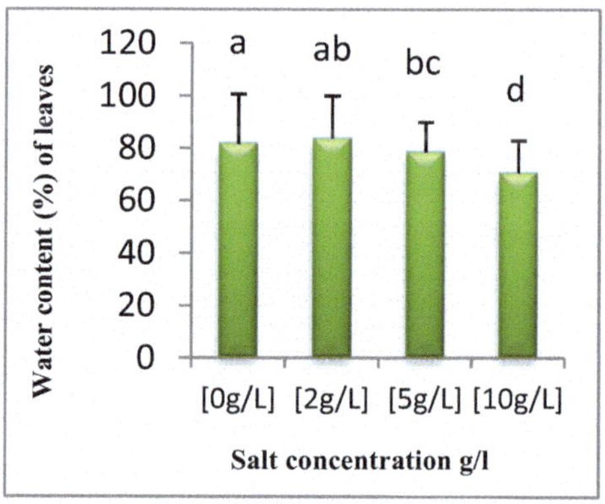

Fig. 12 Variation in water content of sage leaves after 20 days of salt treatment (Values with different letters are significantly different: $p < 0.05$)

Fig. 13 Variation in water content of sage leaves under the effect of different long-term NaCl concentrations in the presence of halophytic plants. (Values with different letters are significantly different: $p < 0.05$)

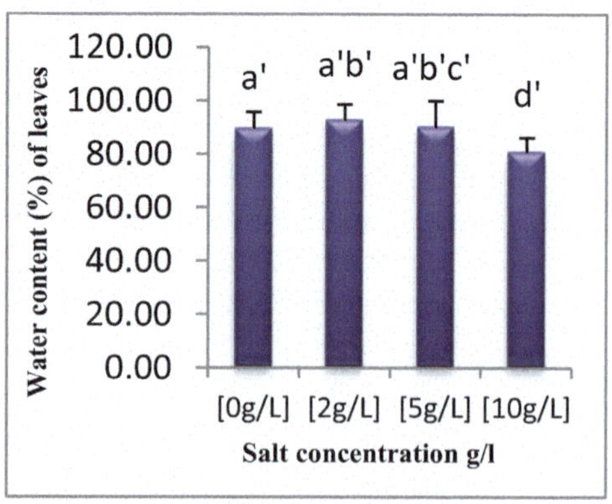

3.6 *Effect of Salt Stress on the Water Content of Sage Stems in the Presence and Absence of* Spergularia Maritima *(Halophyte Plant)*

The results of the effect of salt stress on the water content of sage stems in the presence and absence of *Spergularia maritima* (halophyte plant) show a slight increase in this content in plants subjected to treatments compared with control plants. However, statistical analysis revealed no significant difference between control and stressed treatments, indicating that salt stress has no major effect on the water content of sage stems (Figs. 14 and 15).

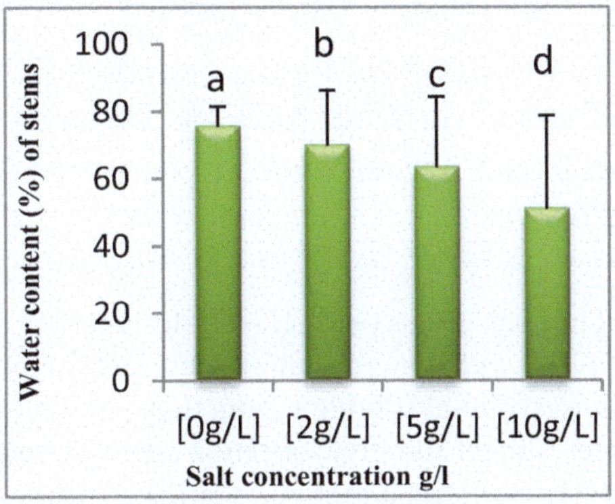

Fig. 14 Variation in water content of sage stems after 20 days of salt treatment. (Values with different letters are significantly different: $p < 0.05$)

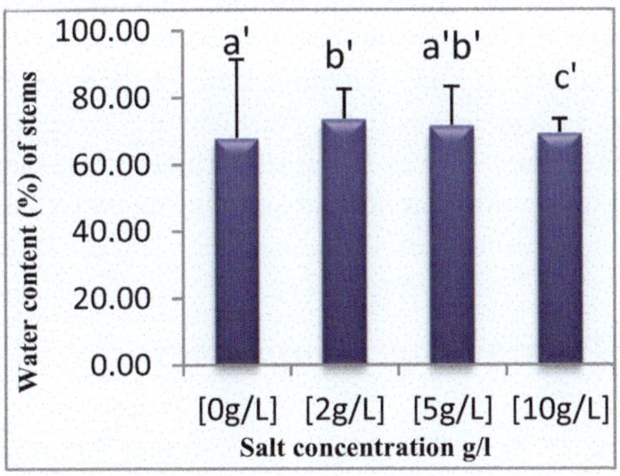

Fig. 15 Variation in water content of sage stems under the effect of different long-term NaCl concentrations in the presence of halophytic plants (Values with different letters are significantly different: $p < 0.05$)

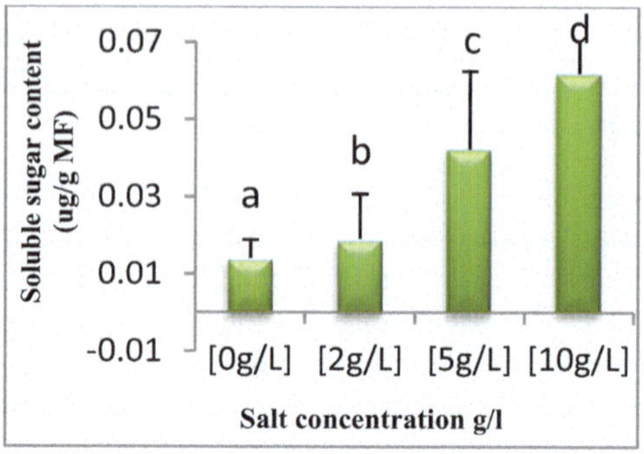

Fig. 16 Evolution of total soluble sugar content in sage leaves as a function of NaCl concentrations (Values with different letters are significantly different: $p < 0.05$)

3.7 *Effect of Salt Stress on Total Soluble Sugars in Sage in the Presence and Absence of* Spergularia Maritima *(Halophyte Plant)*

The results of the effect of salt stress on total soluble sugars in sage in the presence and absence of *Spergularia maritima* (a halophyte plant) show that control plants have a lower accumulation of total soluble sugars than salt-stressed plants. As NaCl concentration increased, so did the concentration of total soluble sugars in sage leaves. Statistical analysis confirms a highly significant difference between control and stressed treatments, indicating that salt stress has a significant effect on the accumulation of total soluble sugars in sage leaves.

By examining Fig. 16, it can be observed that the concentration of total soluble sugars in sage leaves increases gradually with the increase in NaCl concentration. The differences observed between the concentrations are statistically significant. Similarly, the results shown in Fig. 17 demonstrate an increase in the concentration of total soluble sugars in sage leaves in the presence of *Spergularia maritima*, as the NaCl concentration increases. However, it is important to note that at a concentration of 10 g/l of NaCl, the concentration of total soluble sugars remains constant.

3.8 *Effect of Salinity on Sage Morphology*

The results from Figs. 18 and 19 highlight the survival duration of the leaves and stems of the plant under salt stress. Both the control plants and those irrigated with a concentration of 2 g/l show normal growth throughout the experiment, from the beginning of the treatment until the end. However, for the 5 g/l concentration, there is normal growth in the initial days but a slight decrease starting from the 13th day. For the 10 g/l concentration, there is a rapid decline in growth starting from the 6th

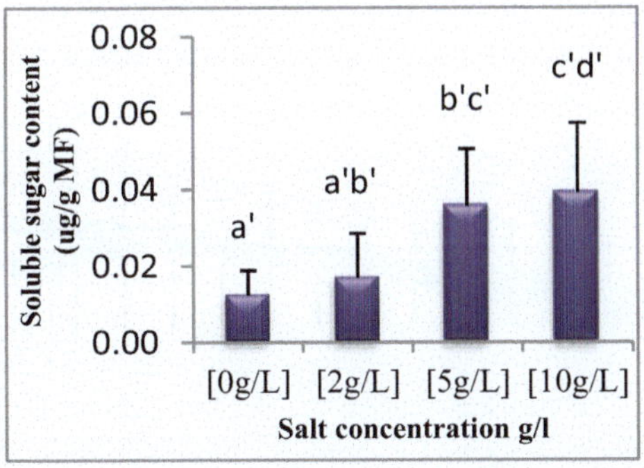

Fig. 17 Changes in total soluble sugars in sage leaves in the presence of halophytic plants as a function of NaCl concentrations (values with different letters are significantly different: $p < 0.05$)

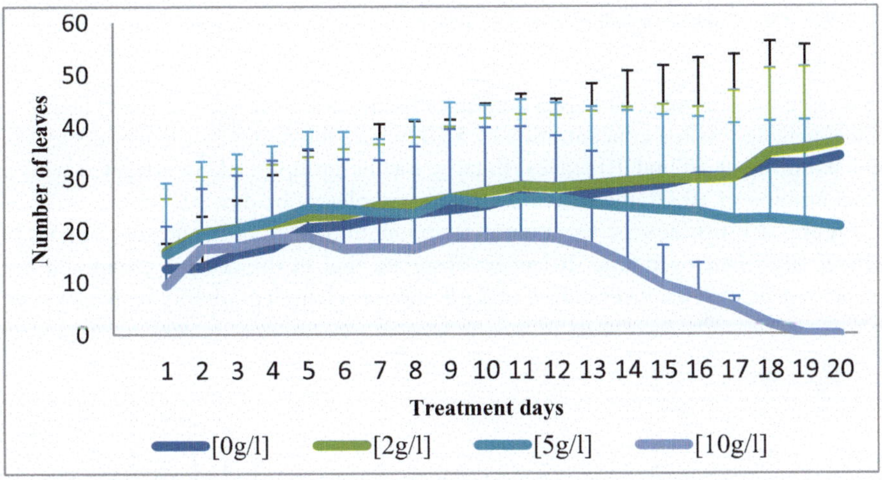

Fig. 18 Effect of long-term salt treatment on sage stem length as a function of time

day of treatment. It should be noted that the survival duration of the sage is 20 days, 13 days, and 6 days for the 2 g/l, 5 g/l, and 10 g/l concentrations, respectively.

3.9 *Effect of Salinity on Sage Morphology in the Presence of* Spergularia Maritima

Figure 20 shows the evolution of the number of sage leaves over time in the presence of *Spergularia maritima.* A slight increase in the number of leaves is observed during the first 14 days for all treatments. The plants then split into two distinct groups. The first group comprises the control group and plants irrigated with a concentration of

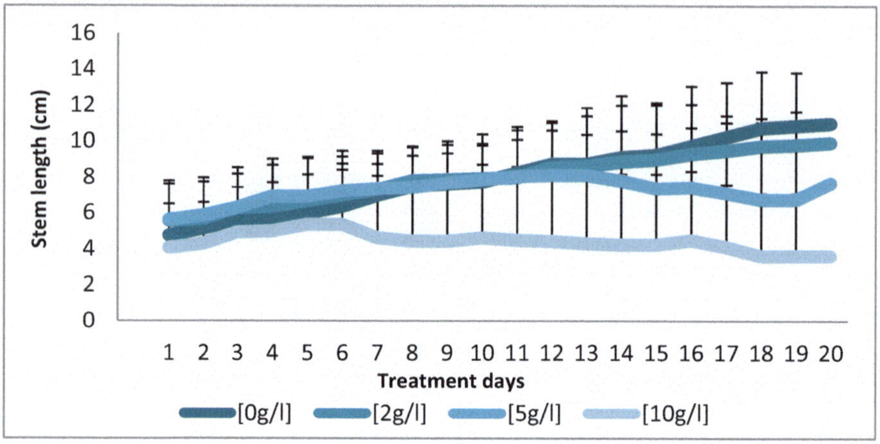

Fig. 19 Effect of long-term salt treatment on sage stem length as a function of time

2 g/l, where the number of leaves continues to increase. The second group includes plants irrigated with a concentration of 5 g/l, where the number of leaves decreases and stabilizes at around 10 leaves. Finally, for the group of plants irrigated with a concentration of 10 g/l, the leaves disappear completely by day 20.

Figure 21 illustrates the variation in stem length of sage in the presence of *Spergularia maritima* over time. It can be observed that in the control group and the group treated with a concentration of 2 g/l, the stem length gradually increases over

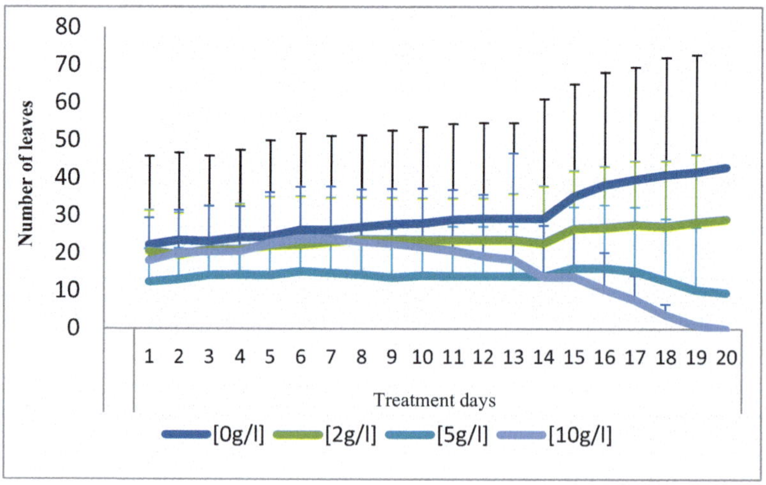

Fig. 20 Effect of long-term salt treatment on the number of sage leaves in the presence of *Spergularia maritima* as a function of time

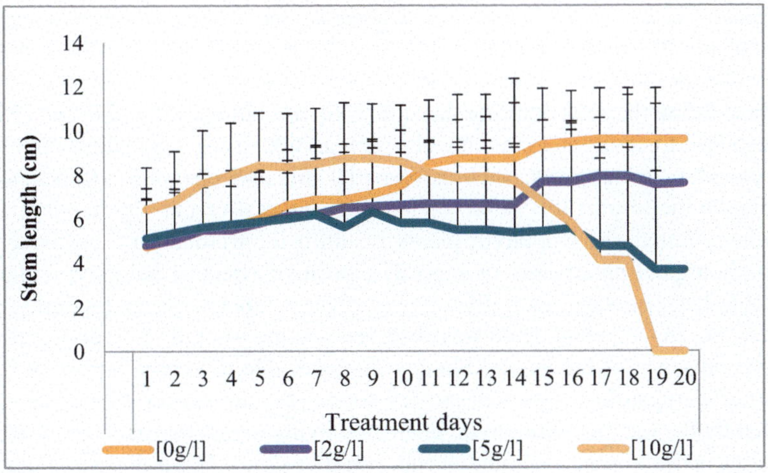

Fig. 21 Effect of long-term salt treatment on stem length (cm) of sage in the presence of *Spergularia maritima* as a function of time

the course of the 20-day experiment. However, for the group treated with a concentration of 5 g/l, the stem length increases during the first 7 days and then slightly decreases by the 20th day. In the group treated with a concentration of 10 g/l, the stem length exhibits significant growth initially, but starts to decrease from the 10th day, reaching 0 cm by the 19th day. It is noteworthy that the survival duration of sage in the presence of *Spergularia maritima* is 20 days, 19 days, and 20 days for the 2 g/l, 5 g/l, and 10 g/l concentrations, respectively.

4 Discussion

The detrimental effects of salinity on plant growth are generally associated with the low osmotic potential of the soil solution and the high toxicity of sodium, which cause multiple disruptions to plant metabolism, growth, and development at the molecular, biochemical, and physiological levels [19–22]. According to the results obtained, salinity further reduced the growth of sage roots. Indeed, root length gradually decreased with increasing intensity of salt stress, as did root volume. This decrease can be attributed to the presence of salt in the growth medium, which limits the plant's calcium supply and leads to inhibition of root and absorptive hair emergence and growth [23]. For stem elongation, it is observed that after 20 days of salt stress application, a significant difference in stem length was observed between the control plants and those subjected to salt treatments. The reduction in aerial growth can be explained by disturbances in the levels of certain growth regulators, particularly abscisic acid and cytokinins induced by salt [24], as confirmed by [25]. The relative water content is a good indicator of water status, and it slightly decreases in salt-stressed plants. The nonsignificant difference in sage water content is due to reduced transpiration and sufficient water uptake by the roots to prevent dehydration of plant tissues and dilute the salts entering the cells [26], as confirmed by [27] and [28].

In response to environmental changes such as salt stress, sage adjusts the levels of its secondary metabolites, such as essential oils and phenolic diterpenes [29]. According to the results, it is observed that sage has a strong ability to adapt to salt stress when planted with halophytic plants, resulting in variations in physicochemical and morphological parameters of sage. The positive effect of halophytic plants is remarkable because sage, as a glycophytic plant, cannot withstand salinity when planted alone. In fact, plant death was observed starting from the 5 g/l concentration, but in the presence of halophytic plants, sage was able to survive up to the 10 g/l concentration, regarding its growth and development in saline conditions. This can be explained by the fact that halophytic plants are living organisms capable of actively functioning in the presence of high salt concentrations [30]. This is the essential aspect of their salt tolerance, which allows them to withstand salt concentrations ranging from 500 mM to 1 M, by enhancing the efficiency of mechanisms for managing the toxicity of the sodium ion Na^+ [31].

5 Conclusion

In conclusion, saline stress has negative effects on the growth of sage when planted alone, resulting in decreased growth of its aerial and root parts, reduced water content (WC), and increased accumulation of total soluble sugars (TSS). However, these effects were less pronounced when sage was planted with *Spergularia maritima* in the same environment, leading to growth and increased development of different sage parts. This study highlights the importance of understanding and harnessing the interactions between halophytic and glycophytic plants to address the challenges posed by climate change and soil salinity. These findings pave the way for innovative agronomic strategies to enhance crop productivity in saline environments while ensuring sustainable use of natural resources. It is worth emphasizing that this study holds particular significance in the context of climate change, which is leading to increased soil salinity and growing water scarcity. The results underscore the potential for domesticating sage as a major crop in saline lands through intercropping with halophytic plants. Considering these results, it is evident that the association between halophytic and glycophytic plants offers significant benefits. By reducing soil salinity and promoting crop growth, this approach opens up new prospects for more efficient agricultural exploitation, both in Morocco and worldwide.

References

1. Dell, M., Jones, B.F., Olken, B.A.: What do we learn from the weather? The new climate-economy literature. J. Econ. Lit. **52**(3), 740–798 (2014)
2. Niu, C.D., Jiang, L.j., Cao, F.G., Liu, C., Guo, J.X., Zhang, Z.T., Yue, Q.Y., Hou, N., Liu, Z. Y., Li, X.W., Tahir, M.M., He, J.Q., Li, Z.X., Li, C., Ma, F.W., Guan, Q.M.: Methylation of a MITE insertion in the MdRFNR1–1 promoter is positively associated with its allelic expression in apple in response to drought stress. The Plant Cell, **34**(10), 3983–4006 (2022)
3. Gokçe, A.F., Chaudhry, U.K.: Use of QTL in developing stress tolerance in agronomic crops, pp. 527–556. Agronomic Crops Springer, Singapore (2020)
4. Zhou, Z.X., Li, Z.Y., Zhang, Z.Q., You, L.R., Xu, L.F., Huang, H.Y., Wang, X.P., Gao, Y., Cui, X.J.: Treatment of the saline-alkali soil with acidic corn stalk biochar and its effect on the sorghum yield in western Songnen Plain. Sci. Total. Environ. **797**, 1–9 (2021)
5. Li, H., Zhao, Q.Y., Huang, H.: Current states and challenges of salt-affected soil remediation by cyanobacteria. Sci. Total. Environ. **669**, 258–272 (2019)
6. Tomaz, A., Palma, P., Alvarenga, P., Gonçalves, M.C.: Soil salinity risk in a climate change scenario and its effect on crop yield. Clim. Chang. Soil Interact. **13**, 351–396 (2020)
7. Li, H., La, S., Zhang, X., Gao, L., Tian, Y.: Salt-induced recruitment of specific root-associated bacterial consortium capable of enhancing plant adaptability to salt stress. ISME J. **15**, 2865–2882 (2022)
8. Zhang, Q., Zhou, W., Liang, G.Q., Sun, J.W., Wang, X.B., He, P.: Distribution des nutriments du sol, des activités enzymatiques extracellulaires et des communautés microbiennes à travers les fractions granulométriques dans une expérience d'engrais à long terme. Appl. Écol. **94**, 59–71 (2015)
9. Hanin, M., Ebel, C., Ngom, M., Laplaze,.L., Masmoudi, K.: New insights on plant salt tolerance mechanisms and their potential use for breeding. Front. Plant Sci. **7**, 1787 (2016)
10. Patel,M.K., Kumar, M., Li,W., Luo,Y., Burritt,D.J., Alkan,N.,TranL.S.P.: Enhancing salt tolerance of plants: from metabolic reprogramming to exogenous chemical treatments and molecular approaches. Cells, **9** (11), Article 2492, https://doi.org/10.3390/cells9112492 (2020)
11. Zörb, C., Geilfus,C.M.,Dietz, K.J.: Salinity and crop yield. Plant Biol. 31–38 (2019)
12. Gupta, B., Huang, B.: Mechanism of salinity tolerance in plants: physiological, biochemical, and molecular characterization. Int. J. Genom. pp. 1–18 (2014)
13. Megdiche, W., Amor, N.B., Debez, A., Hessini, K., Ksouri, R., Zuily-Fodil, Y., Abdelly, C.: Salt tolerance of the annual halophyte *Cakile maritima* as affected by the provenance and the developmental stage. Acta Physiol. Plant. **29**, 375–384 (2007)
14. Sabry, M.M., Abdel-Rahman, R.F., El-Shenawy, S.M., Hassan, A.M., El-Gayed, S.H.: Estrogenic Activity of Sage (*Salvia Officinalis* L.) Aerial Parts and Its Isolated Ferulic Acid in Immature Ovariectomized Female Rats. J. Ethnopharmacol. 282 (janvier): 114579 (2022)
15. Hedge, I.C.: A global survey of the biogeography of the Labiatae. Royal Botanical Gardens, pp. 7–17. Kew (1992)
16. Gerstberger, P.: Die Salz-Schuppenmiere (*Spergularia salina*) als Besiedlerse kundärer Salzstandorte in Bayern, no 12: 361–65. Tuexenia (1992)
17. Musick, G. J., Fairchild, M. L., Fergason,V. L., Zuber, M. S.: A method of measuring root volume in corn (*Zea mays* L.). Crop Science 5 (6), 601–2 (1965)
18. DuBois, M., Gilles, K.A., Hamilton, J.K., Rebers, P.A.T., Smith, F.: Colorimetric method for determination of sugars and related substances. Anal. Chem. **28**(3), 350–356 (1956)
19. Winicov, I.: New molecular approaches to improving salt tolerance in crop plants. Ann. Bot, **82**(6): 703–710. Lond (1998)
20. Munns, R.: Comparative physiology of salt and water stress. Plant Cell Environ **25**(2): 239–250. PMID:11841667 (2002)
21. Tester, M., Davenport, R.: Na+ tolerance and Na+ transport in higher plants. Ann. Bot. **91**(5), 503–527 (2003)
22. Yamaguchi-Shinozaki, K., Shinozaki, K.: Organization of cisacting regulatory elements in osmotic- and cold-stress-responsive promoters. Trends Plant Sci. **10**(2), 88–94 (2005)

23. Zahran, H., Sprent, I.: Effet of sodium chloride and polyethylene glycol on root- hair infection and nodulation of viciafaba L. plant by Rhizobium leguminosarum. Planta 167, 303–309 (1986)
24. Kuiper, D., Schuit, J., Kuiper, P.J.C.: Actual cytokinin concentrations in plant tissue as an indicator for salt resistance in cereals. Plant Soil **123**, 243–250 (1990)
25. Termaat, A., Passora, J.B., Munns, R.: Shoot turgor does not limit shoot growth of NaCl affected wheat and barley. Plant Physiol. **77**, 869–872 (1985)
26. Ngasamy, P.: World salinization with emphasis on Australia. J. Exp. Botany **57**, 1017, 1023 (2006)
27. Hasegawa, P. M., Bressan, R. A., Zhu, J. K., Bohnert, H. J.: Plant cellular and molecular responses to high salinity. Edit. Plant Mol. Biol. 51, 499 (2000)
28. Fricke, W., Peters, W. S.: The biophysics of leaf growth in salt stressed barley. A study at the cell level. Plant Physiol. 129, 388 (2002)
29. Tounekti, T., Hernandez, I., Muller, M., Khemira, H., Munne-Bosch, S.: Kinetin applications alleviate salt stress and improve the antioxidant composition of leaf extracts in Salvia officinalis. Plant Physiol. Biochem. **49**(10), 1165–1176 (2011)
30. Hopkins, W.G.: Physiologie végétale. Traduction de la 2e édition américaine. Ed. de Boeck. Bruxelles (2003)
31. Flowers, T.J., Galal, H.K., Bromham, L.: Evolution of halophytes: multiple origins of salt tolerance in land plants. Funct. Plant Biol. **37**, 604–612 (2010)

Phytoremediation Processes for the Removal of Heavy Metals

Khadija El-Moustaqim, Anas Tallou, Jamal Mabrouki, Miloudia Slaoui, and Driss Hmouni

Abstract The development of human activities, whether domestic or industrial, depends on a water source. After multiple uses, water quality deteriorates, resulting in the production of wastewater. The growth in the number of nutrients in wastewater and their adverse effects on the environment have sparked interest in developing more efficient and profitable approaches to wastewater treatment as solutions to conventional methods. Microalgae are a group of microorganisms that can colonize almost any ecosystem on Earth. Most of these microorganisms are autotrophic, transforming light and nutrients into biomass through photosynthesis. The most widely-used method Phytoremediation of wastewater using micro-algae offers great benefits in terms of nutrient removal efficiency and low cost. In micro-algal-based bio-remediation, microalgae bind carbon dioxide and generate oxygen through the process of photosynthesis, thereby increasing the biological oxygen demand of the polluted water. The aim of the current review is to examine the impact of microalgae on the treatment of heavy metals in wastewater and to see how effective this new strategy is in improving biological remediation so that it can adapt to changing environmental conditions.

Keywords Heavy metals · Biodegradation · Phytoremediation · Wastewater

---

K. El-Moustaqim (✉) · D. Hmouni
Improvement and Valuation of Plant Resources, Faculty of Sciences, Ibn Tofaïl University—KENITRA-University Campus, 14000 Kenitra, Morocco
e-mail: elmoustaqim.khadija@gmail.com

A. Tallou
Department of Soil, Plant and Food Sciences, University of Bari "Aldo Moro", Bari, Italy

J. Mabrouki
Laboratory of Spectroscopy, Molecular Modelling, Materials, Nanomaterials, Water and Environment, CERN2D, Mohammed V University in Rabat, Faculty of Science, AV Ibn Battouta, Agdal, 10106 Rabat, Morocco

M. Slaoui
Energy, Materials and Sustainable Development (EMDD) Laboratory-Higher School of Technology-SALE, Center of Water, Natural Resources, Environment and Sustainable Development (CERN2D), University Mohammed V in RABAT, 10100 Rabat, Morocco

J. Mabrouki and M. Azrour (eds.), *Integrated Solutions for Smart and Sustainable Environmental Conservation*, Studies in Systems, Decision and Control 527,
https://doi.org/10.1007/978-3-031-55787-3_3

1 Introduction

The impact of disease on the world's population is automatically linked to its impact on water, and disease transmission through polluted water does not stop in developing countries [1, 2]. Due to fast urban industrialization and economic development, zones continue to accumulate in coastal environments as a result of effluent from rivers, wastewater, as well as wind power extraction methods [3]. Thus it undoubtedly brings pollution of the natural environment which is greatly increased [4–6]. It may be noted that the imbalance of availability and demand is the subject of water scarcity, which is of course caused by the rapid growth of the urban population, the discontinuous development of the water reserve, periods of severe surface water drought, and groundwater pollution [7–9]. The trace metals are not detectable by any biological or physical process and persist in the soils for a long-term period of time, posing a very long-term menace to the ecosystem [3, 10]. Among other things, copper and zinc also pose a major threat to public health, due to their persistence and accumulation in food chains [11]. Depending on the role they play in the biological process, heavy metals can be classed as essential and inessential [12]. Of the different heavy metals, copper (Cu) and zinc (Zn) are the most organic, the most exchanged, and the most water-soluble and, therefore, are easily liberated by sediments into the overlying water bodies [13]. However, they can become toxic if they are too abundant. Heavy metals such as plumbism, cadmium, arsenic, and mercury are very toxic and have no known plant function [14]. They represent a serious risk to human health due to their penetration into the human body through biomagnification [15, 16]. Microalgae is an extensively studied organism group for their potential in ecological remediation [17, 18]. The idea of making use of plants to reduce environmental contamination has been around for a long time [19–22]. The process of phytoremediation consists of using plants to remove contaminants by destroying, inactivating, or immobilizing them in non-hazardous states [23, 24]. As well as has great value for its high efficiency in nutrient removal and its low cost-effectiveness [25, 26]. The principal objective is to provide some kind of overview of recent advances in the development of phytoremediation techniques by microalgae, including strategies to enhance the bioavailability, tolerance, accumulation, as well as efficiency of microalgae in the removal of heavy metals.

2 General Principles of Wastewater Composition and Processing

A modification in the intensity and quantity from a whole various series of streams is presented during their contamination in streams. Current technology and societal lifestyles are considered to represent the nature of wastewater [27]. The latter is characterized by a varied mixture of ingredients that have an organic and inorganic

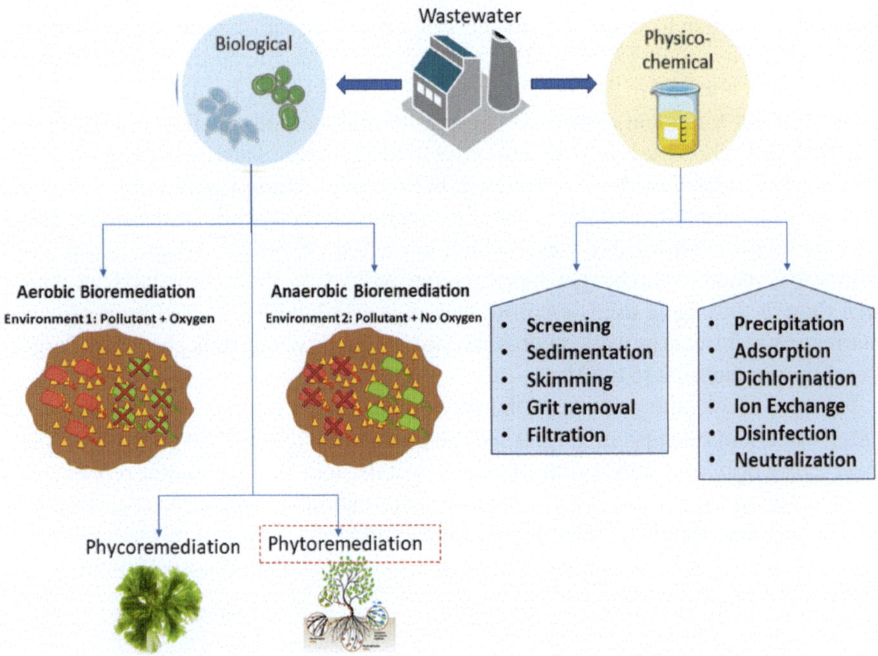

Fig. 1 Representation of the various wastewater treatment methods

structure, plus synthetic substances. The main contaminants in the effluent are carbohydrates, lipids, sucrose, and amino acids which make up three-quarters of the organic carbon found in wastewater [28, 29]. Some variations in inorganic contaminants in wastewater include sodium, calcium, magnesium, potassium, sulphur, arsenic, phosphorus, bicarbon, ammonium salts, and heavy metals [30–33]. Lead persistence (Pb) can range from 150 to 5000 years in soil and after sludge application to soil, remains in high concentrations up to 150 years [34, 35].

Inefficient water management strategies mean that the treatment of wastewater plays a crucial role in ensuring that harmful contaminants are removed from wastewater before it is discharged into recipient water sources or used again for useful processes [36–38]. Figure 1 is a diagram that summarizes the different methods of wastewater treatment. Conventional systems include chemical treatment, activated sludge treatment, anaerobic digestion, membrane filtration, and more.

3 Dimensions of the Phytoremediation Process

The principle behind the phytoremediation method is based on the application of plants that can develop in a contaminated environment and that make a most interesting contribution to influencing biological processes, chemical, as well as physical, that occur in order to contribute to the effective removal of xenobiotics from the biological environment [39–44]. This technology is recognized as a useful wastewater treatment alternative that is being researched worldwide. It offers great potential for wastewater treatment compared to conventional methods such as run-off filters, batch reactors, and others [45–47]. In order to treat contaminated sites, several remediation techniques have been developed, but the application is valid only for some of them for the treatment of heavy metal contamination [48–50]. Durability of metal waste is longer than organic waste which has a fast character in degradation. Poor solid waste management by developing countries leads to environmental degradation even though they have used scientific techniques in landfills caused by pesticides, pharmaceutical waste as well as personal care products, which took care of the largest proportion in landfills [51–53].

Figure 2 shows the process of the phytoremediation method. This technique is widely used and most promising, since the plants used and for the purpose of tolerance, extraction, and absorption of heavy metals, and this technique represents a positive character that includes several aspects including cost-effective, economically sound, and widely adopted by researchers [54, 55]. And that this technology is applicable in several contaminated sites with lower installation costs [56–58].

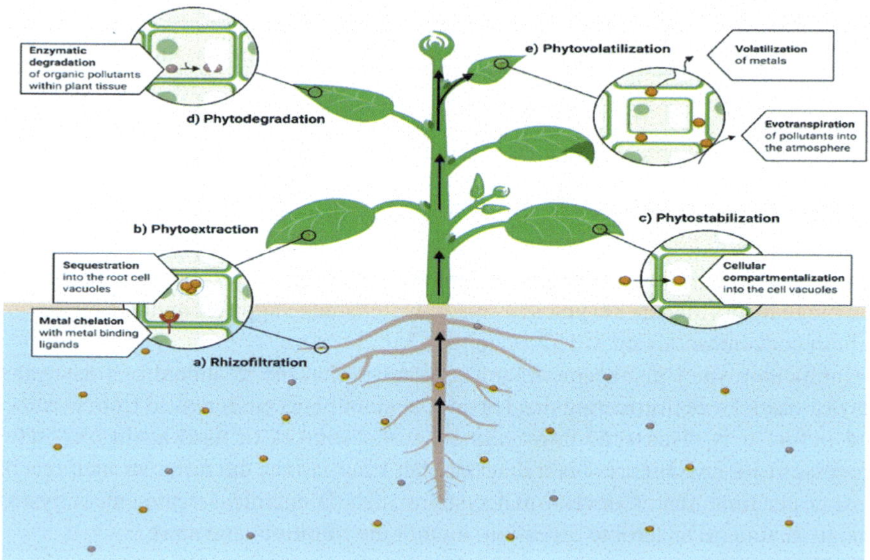

Fig. 2 Phytoremediation processes in wastewater

4 Phytoremediation as a Bioremediation Strategy

Bioremediation has no harmful influence on the environment, which is based on plants and also on microorganisms such as algae and bacteria [59, 60]. In the context of ensuring respect for the environment, biological sanitation is recognized as an eco-friendly technique as well as being economical compared to a number of other methods explored whether physical or chemical [61, 62]. In addition, microorganisms are found to be highly adapted to phytoremediation treatment based on their development in biological activities [63, 64]. A method of in situ remediation, phytoremediation involves placing plants in contaminated environments to remove, degrade and detoxify toxic metal contamination [65, 66]. It is considered to be the most suitable technique for the lowest levels of contamination and for a longer period without damage to the site [67, 68].

4.1 *Study of the SWOT Analysis of Phytoremediation*

Based on the system's internal and external characteristics that provide support for the operational decision, the following SWOT Analysis is divided into four parts: strengths, weaknesses, opportunities, and threats [32, 69–75]. Figure 3 shows the SWOT analysis of phytoremediation.

In these cases, it is necessary to carry out several experimental studies and a thorough optimization in order to evaluate the usefulness of different types of plants used in phytoremediation [76, 77], and to carry out additional large-scale analyses to confirm performance under different environmental situations that can be directly reproduced in actual field applications [78–80].

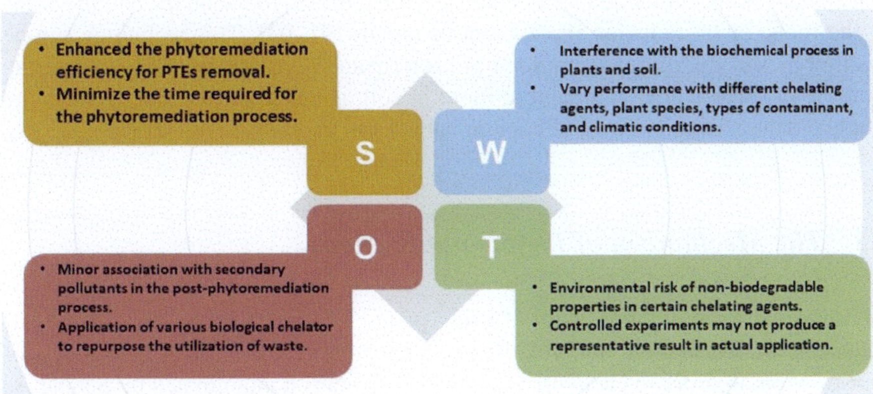

Fig. 3 SWOT analysis of phytoremediation

5 Mechanisms of Phytoremediation of Heavy Metals

Currently, the term phytoremediation is currently used as a reference for various techniques using superior plants for the purpose of eliminating contamination, and this technique includes a mechanism consisting of four parts including phytovolatilization, phytostabilization, phytoextraction, and rhizofiltration [81–84].

5.1 The Mechanism of Phytoextraction

The mechanism is based on the adsorption of pollutants found in soil, water, or sediment [85–87]. The general principle of this technique consists in growing suitable plant varieties in situ, extracting the biomass containing heavy metals, treating it, and reducing its mass, and this can be done by composting, compression, dehydration, and thermal decomposition. Re-extraction of trace elements or disposal as waste from biomass harvested during processing is considered economical for the generation of metallic contaminants [10, 88, 89]. Phytoextraction involves the use of chemicals that are intended to enhance the adsorption of minerals by plants for example Ethylenediaminetetraacetic acid (EDTA), N-hydroxyethyl EDTA (HEDTA), or citric acid [90, 91].

5.2 The Mechanism of Phytostabilization

A process that does not eliminate contaminants months instead, it holds them in the roots by accumulation and absorption or precipitation in the rhizosphere. It can be done by up taking contaminants and their sequestration in roots, on the cellular walls of the roots, or by reducing the content of metals [91–93]. This technique can be ensured by soil microorganisms such as bacteria and mycorrhizae that can have an increase in surface area and capture more in soil, act as a barrier that protects the plant, and the position of ions of contaminants from the roots to the growth [94, 95].

5.3 The Mechanism of Phytovolatilization

Phytovolatilization is the process of using plants to capture heavy metal contaminants and transform them into volatile chemical compounds through the process of transpiration, in order to clean up aquatic environments contaminated by Se and As [96, 97]. Hg can also be eliminated and is considered a neurodegenerative heavy metal that constitutes a serious menace to public health in the food chains [98, 99]. This method is considered to be quite risky since pollutants are penetrated into the atmosphere during their removal, even if their compositions are less toxic, but it still remains the presence of a threat to the ecosystem [40].

5.4 *The Mechanism of Rhizofiltration*

This mechanism relies on the power of the plant's roots to capture as well as sequester metal pollutants in water, which allows reducing metals including Cd, Cr, Cu, Ni, and others [100, 101]. The principle is essentially focused on the roots of species of plants that can capture pollutants from ground and surface water, industrial, agricultural, and other wastewater [102]. This process can be performed on root tissues that do not require an active biological system [103]. Plants exploited in this process are found in a natural aquatic environment that is of interest, however, one finds a type of plants such as those grown in hydroponics or aeroponics have the power to remove contaminants more easily [104].

6 Potential of Phytoremediation for Heavy Metal Elimination

Worldwide, metal ions are becoming a major concern due to their contamination of the ecosystem, which amounts to rapid urbanization and industrialization. They can enter the food chain during their penetration by a living organism, which creates serious dangers [105]. The process of phytoremediation must improve over time in order to increase its performance and yield. Studies continue in research on new techniques such as phytoremediation by microorganisms or by biochar which considers itself an economic product that has a porous character and has a very significant capacity in several contexts [106]. This mechanism appears to be an effective means of removing metals from polluted sites, and that the cultivation of genetically engineered plants has a metal accumulation capacity based on adsorption processes [107]. Protection of sanitation systems may be based on plant tolerance to stress from heavy metals for the purpose of protection and relevant efficacy.

7 Conclusion

The elimination of pollutants remains a major preoccupation that must be addressed immediately, and which represents a problem for both mankind and the ecosystem. Due to their hazardous effects and quick accumulation in the environment, a number of methods are available to prevent or mitigate soil contamination by heavy metals. However, phytoremediation proved to be the most effective method for soil safety and the reduction of heavy metal contamination, which is acceptable to the public and offers numerous advantages. The improved genetic engineering approach is a tool being developed to establish the goal of modifying plants with desirable characteristics, including rapid growth, increased biomass production, and enhanced heavy metal accumulation with adaptation to various geological or climatic situations.

References

1. Elfanssi, S., Mabrouki, J., Oumlouki, K. E., Ghizlane Fattah, Mandi, L.: Modeling and simulation of phytoremediation technology by artificial neural network, in advanced technology for smart environment and energy. In: Mabrouki, J., Mourade, A., Irshad, A., Chaudhry, S. A.: Éd., Environmental science and engineering. Cham: Springer International Publishing, 2023, pp. 83–91. https://doi.org/10.1007/978-3-031-25662-2_7
2. Jamal, M.: Studying the effects of organic and hydraulic shock loads on the membrane bioreactor (MBR) by using GPS-X mathematical model. Int. J. Membr. Sci. Technol. 9, 33–41, 2022, https://doi.org/10.15379/2410-1869.2022.04
3. Kwon, H. K., Jeon, J. Y., Oh, S. J.: Potential for heavy metal (copper and zinc) removal from contaminated marine sediments using microalgae and light emitting diodes. Ocean Sci. J. **52**(1), 57–66 (2017), https://doi.org/10.1007/s12601-017-0001-z
4. Li, M., et al.: Phytochelatin synthase de-regulation in *Marchantia polymorpha* indicates cadmium detoxification as its primary ancestral function in land plants and provides a novel visual bioindicator for detection of this metal. J. Hazard. Mater. **440**, 129844 (2022). https://doi.org/10.1016/j.jhazmat.2022.129844
5. Yan, A., Wang, Y., Tan, S. N., Mohd Yusof, M. L., Ghosh, S., Chen, Z.: Phytoremediation: a promising approach for revegetation of heavy metal-polluted land. Front. Plant Sci. **11** (2020). Consulté le: 18 juin 2023. [En ligne]. Disponible sur: https://www.frontiersin.org/articles/10.3389/fpls.2020.00359
6. Haider, F.U., et al.: Cadmium toxicity in plants: impacts and remediation strategies. Ecotoxicol. Environ. Saf. **211**, 111887 (2021). https://doi.org/10.1016/j.ecoenv.2020.111887
7. Pérez-Mora, L. S. et al.: Phycoremediation processes for secondary effluent from sewage treatment plants using photosynthetic microorganisms: a review. Appl. Microbiol. 3(2), 400–416 (2023). https://doi.org/10.3390/applmicrobiol3020028
8. O. US EPA, Guidelines for Water Reuse, 13 août 2019. https://www.epa.gov/waterreuse/guidelines-water-reuse. consulté le 18 juin 2023
9. Wiener, M. J., Jafvert, C. T., Nies, L. F.: The assessment of water use and reuse through reported data: a US case study. Sci. Total Environ. **539**, 70–77 (2016). https://doi.org/10.1016/j.scitotenv.2015.08.114
10. Suman, J., Uhlik, O., Viktorova, J., Macek, T.: Phytoextraction of heavy metals: a promising tool for clean-up of polluted environment? Front. Plant Sci. **9** (2018). Consulté le: 18 juin 2023. [En ligne]. Disponible sur: https://doi.org/10.3389/fpls.2018.01476
11. Choi, K. Y., Kim, S. H., Hong, G. H., Chon, H. T.: Distributions of heavy metals in the sediments of South Korean harbors. Environ. Geochem. Health **34**, 71–82 (2012)
12. Ashraf, S., Ali, Q., Zahir, Z. A., Ashraf, S., Asghar, H. N.: Phytoremediation: environmentally sustainable way for reclamation of heavy metal polluted soils. Ecotoxicol. Environ. Saf. **174**, 714–727 (2019). https://doi.org/10.1016/j.ecoenv.2019.02.068
13. Han, B.-C., Jeng, W.-L., Hung, T.-C., Ling, Y.-C., Shieh, M.-J., Chien, L.-C.: Estimation of metal and organochlorine pesticide exposures and potential health threat by consumption of oysters in Taiwan. Environ. Pollut. **109**(1), 147–156 (2000)
14. Fasani, E., Manara, A., Martini, F., Furini, A., DalCorso, G.: The potential of genetic engineering of plants for the remediation of soils contaminated with heavy metals. Plant Cell Environ. **41**(5), 1201–1232 (2018). https://doi.org/10.1111/pce.12963

15. ur Rehman, M. Z., et al.: Remediation of heavy metal contaminated soils by using *Solanum nigrum*: a review. Ecotoxicol. Environ. Saf. **143**, 236–248 (2017). https://doi.org/10.1016/j.ecoenv.2017.05.038
16. Sarwar, N., et al.: Phytoremediation strategies for soils contaminated with heavy metals: modifications and future perspectives. Chemosphere **171**, 710–721 (2017). https://doi.org/10.1016/j.chemosphere.2016.12.116
17. Lim, S.-L., Chu, W.-L., Phang, S.-M.: Use of *Chlorella vulgaris* for bioremediation of textile wastewater. Bioresour. Technol. 101(19), 7314–7322 (2010)
18. Tsuji, N., et al.: Regulation of phytochelatin synthesis by zinc and cadmium in marine green alga *Dunaliella tertiolecta*. Phytochemistry **62**(3), 453–459 (2003)
19. Vats, P., Kaur, U. J., Rishi, P.: Heavy metal-induced selection and proliferation of antibiotic resistance: a review. J. Appl. Microbiol. **132**(6), 4058–4076 (2022). https://doi.org/10.1111/jam.15492
20. Ghori, N.-H., et al.: Heavy metal stress and responses in plants. Int. J. Environ. Sci. Technol. 16(3), 1807–1828 (2019). https://doi.org/10.1007/s13762-019-02215-8
21. Kosakivska, I. V., Babenko, L. M., Romanenko, K. O., Korotka, I. Y., Potters, G.: Molecular mechanisms of plant adaptive responses to heavy metals stress. Cell Biol. Int. 45(2), 258–272 (2021). https://doi.org/10.1002/cbin.11503
22. Choudhary, S., Wani, K. I., Naeem, M., Khan, M. M. A., Aftab, T.: Cellular responses, osmotic adjustments, and role of osmolytes in providing salt stress resilience in higher plants: polyamines and nitric oxide crosstalk. J. Plant Growth Regul. **42**(2), 539–553 (2023). https://doi.org/10.1007/s00344-022-10584-7
23. Abdel-Shafy, H. I., Mansour, M. S. M.: Phytoremediation for the elimination of metals, pesticides, PAHs, and other pollutants from wastewater and soil. In: Kumar, V., Kumar, M., Prasad, R.: Éd., Phytobiont and ecosystem restitution. Singapore: Springer, 2018, pp. 101–136. https://doi.org/10.1007/978-981-13-1187-1_5
24. Dewinta, A. F., Hartono, E. S., Yusni, E., Susetya, I. E., Siregar, R. F.: The ability of spirulina sp. microalgae as a phytoremediation agents in liquid waste of handling fish from cemara market. In: Medan, J. Ilm. Perikan. Dan Kelaut. 12(2), 286–295 (2020). https://doi.org/10.20473/jipk.v12i2.22224
25. Nie, X., Mubashar, M., Zhang, S., Qin, Y., Zhang, X.: Current progress, challenges and perspectives in microalgae-based nutrient removal for aquaculture waste: a comprehensive review. J. Clean. Prod. 277, 124209 (2020)
26. Mocek-Płóciniak, A., Mencel, J., Zakrzewski, W., Roszkowski, S.: Phytoremediation as an effective remedy for removing trace elements from ecosystems. Plants 12(8), 1653 (2023). https://doi.org/10.3390/plants12081653
27. Abdel-Raouf, N., Al-Homaidan, A. A., Ibraheem, I. B. M.: Microalgae and wastewater treatment. Saudi J. Biol. Sci. 19(3), 257–275 (2012). https://doi.org/10.1016/j.sjbs.2012.04.005
28. Liu, Z.-Q. et al.: Activated carbon catalytic ozonation of reverse osmosis concentrate after coagulation pretreatment from coal gasification wastewater reclamation for zero liquid discharge, J. Clean. Prod. 286, 124951 (2021). https://doi.org/10.1016/j.jclepro.2020.124951
29. Hasan, R.: Bioremediation of swine wastewater and biofuel potential by using *Chlorella vulgaris*, *Chlamydomonas reinhardtii*, and *Chlamydomonas debaryana*. J. Pet. Environ. Biotechnol. **05**(03) (2014). https://doi.org/10.4172/2157-7463.1000175
30. Zhao, K., Ge, L., Wong, T. I., Zhou, X., Lisak, G.: Gold–silver nanoparticles modified electrochemical sensor array for simultaneous determination of chromium(III) and chromium(VI) in wastewater samples. Chemosphere 281, 130880 (2021). https://doi.org/10.1016/j.chemosphere.2021.130880
31. Alengebawy, A., Abdelkhalek, S. T., Qureshi, S. R., Wang, M.-Q.: Heavy metals and pesticides toxicity in agricultural soil and plants: ecological risks and human health implications. Toxics 9(3), Art. no 3 (2021). https://doi.org/10.3390/toxics9030042
32. Yang, Q., Yang, C., Yu, H., Zhao, Z., Bai, Z.: The addition of degradable chelating agents enhances maize phytoremediation efficiency in Cd-contaminated soils. Chemosphere 269, 129373 (2021). https://doi.org/10.1016/j.chemosphere.2020.129373

33. Ahamed, A., Ge, L., Zhao, K., Veksha, A., Bobacka, J., Lisak, G.: Environmental footprint of voltammetric sensors based on screen-printed electrodes: An assessment towards "green" sensor manufacturing. Chemosphere 278, 130462 (2021). https://doi.org/10.1016/j.chemosphere.2021.130462
34. Berglund, M. et al.: Exposure determinants of cadmium in European mothers and their children. Environ. Res. 141, 69–76 (2015). https://doi.org/10.1016/j.envres.2014.09.042
35. Jabeen, R., Ahmad, A., Iqbal, M.: Phytoremediation of heavy metals: physiological and molecular mechanisms. Bot. Rev. 75(4), 339–364 (2009). https://doi.org/10.1007/s12229-009-9036-x
36. Lugo, A., Xu, X., Abeysiriwardana-Arachchige, I. S. A., Bandara, G. L. C. L., Nirmalakhandan, N., Xu, P.: Techno-economic assessment of a novel algal-membrane system versus conventional wastewater treatment and advanced potable reuse processes: part II. J. Environ. Manage. 331, 117189 (2023). https://doi.org/10.1016/j.jenvman.2022.117189
37. Krishnan, R. Y., Manikandan, S., Subbaiya, R., Karmegam, N., Kim, W., Govarthanan, M.: Recent approaches and advanced wastewater treatment technologies for mitigating emerging microplastics contamination – a critical review. Sci. Total Environ. 858, 159681 (2023). https://doi.org/10.1016/j.scitotenv.2022.159681
38. Kim, M., Cui, F.: Identification of bacterial communities in conventional wastewater treatment sludge to inform inoculation of the anammox process. Chemosphere 311, 137167 (2023). https://doi.org/10.1016/j.chemosphere.2022.137167
39. Raskin, I., Kumar, P. N., Dushenkov, S., Salt, D. E.: Bioconcentration of heavy metals by plants. Curr. Opin. Biotechnol. 5(3), 285–290 (1994). https://doi.org/10.1016/0958-1669(94)90030-2
40. Meagher, R. B., Heaton, A. C. P.: Strategies for the engineered phytoremediation of toxic element pollution: mercury and arsenic. J. Ind. Microbiol. Biotechnol. 32(11–12), 502–513 (2005). https://doi.org/10.1007/s10295-005-0255-9
41. McCutcheon, S. C., Rock, S. A.: Phytoremediation: state of the science conference and other developments. Int. J. Phytoremediation 3(1), 1–11 (2001). https://doi.org/10.1080/15226510108500047
42. Susarla, S., Medina, V. F., McCutcheon, S. C.: Phytoremediation: an ecological solution to organic chemical contamination. Ecol. Eng. **18**(5), 647–658 (2002). https://doi.org/10.1016/S0925-8574(02)00026-5
43. Băbău, A. M. C., Micle, V., Damian, G. E., Sur, I. M.: Sustainable ecological restoration of sterile dumps using *Robinia pseudoacacia*. Sustainability **13**(24), Art. no 24 (2021). https://doi.org/10.3390/su132414021
44. Pang, Y. L., Quek, Y. Y., Lim, S., Shuit, S. H.: Review on phytoremediation potential of floating aquatic plants for heavy metals: a promising approach. Sustainability 15(2), Art. no 2 (2023). https://doi.org/10.3390/su15021290
45. Srinuanpan, S., Cheirsilp, B., Kassim, M. A.: Chapter 5 – oleaginous microalgae cultivation for biogas upgrading and phytoremediation of wastewater. In: Yousuf, A., Éd., Microalgae cultivation for biofuels production, Academic Press, 2020, pp. 69–82. https://doi.org/10.1016/B978-0-12-817536-1.00005-9
46. Rajhi, H., Bardi, A.: Chapter 4 – Phytoremediation of endocrine disrupting pollutants in industrial wastewater. In: Haq, I., Kalamdhad, A., Pandey, A., Éd., Current developments in biotechnology and bioengineering, Elsevier, pp. 55–84 (2023). https://doi.org/10.1016/B978-0-323-91902-9.00002-X
47. Zulkernain, N. H., Uvarajan, T., Ng, C. C.: Roles and significance of chelating agents for potentially toxic elements (PTEs) phytoremediation in soil: a review. J. Environ. Manage **341**, 117926 (2023). https://doi.org/10.1016/j.jenvman.2023.117926
48. Huang, D., et al.: Nanoscale zero-valent iron assisted phytoremediation of Pb in sediment: impacts on metal accumulation and antioxidative system of *Lolium perenne*. Ecotoxicol. Environ. Saf. **153**, 229–237 (2018). https://doi.org/10.1016/j.ecoenv.2018.01.060
49. Sharma, P., Tripathi, S., Chandra, R.: Phytoremediation potential of heavy metal accumulator plants for waste management in the pulp and paper industry. Heliyon **6**(7) (2020). https://doi.org/10.1016/j.heliyon.2020.e04559

50. Peng, J., Song, Y., Yuan, P., Cui, X., Qiu, G.: The remediation of heavy metals contaminated sediment. J. Hazard. Mater. 161(2), 633–640 (2009). https://doi.org/10.1016/j.jhazmat.2008.04.061
51. Hatamian, M., Noshad, M., Abdanan-Mehdizadeh, S., Barzegar, H.: Effect of roasting treatment on functional and antioxidant properties of chia seed flours. NFS J. 21, 1–8 (2020). https://doi.org/10.1016/j.nfs.2020.07.004
52. Ojuri, O. O., Ayodele, F. O., Oluwatuyi, O. E.: Risk assessment and rehabilitation potential of a millennium city dumpsite in Sub-Saharan Africa. Waste Manag. 76, 621–628 (2018). https://doi.org/10.1016/j.wasman.2018.03.002
53. Hao, Z., et al.: Heat generation and accumulation in municipal solid waste landfills. Environ. Sci. Technol. 51(21), 12434–12442 (2017). https://doi.org/10.1021/acs.est.7b01844
54. Verma, R. L., Borongan, G.: Emissions of greenhouse gases from municipal solid waste management system in Ho Chi Minh city of Viet Nam. Urban Sci. 6(4), Art. no 4 (2022). https://doi.org/10.3390/urbansci6040078
55. Seneviratne, M., Rajakaruna, N., Rizwan, M., Madawala, H. M. S. P., Ok, Y. S., Vithanage, M.: Heavy metal-induced oxidative stress on seed germination and seedling development: a critical review. Environ. Geochem. Health 41(4), 1813–1831 (2019). https://doi.org/10.1007/s10653-017-0005-8
56. Li, Z., et al.: Enhanced phytoremediation of selenium using genetically engineered rice plants. J. Plant Physiol. 271, 153665 (2022). https://doi.org/10.1016/j.jplph.2022.153665
57. Mamirova, A., Baubekova, A., Pidlisnyuk, V., Shadenova, E., Djansugurova, L., Jurjanz, S.: Phytoremediation of soil contaminated by organochlorine pesticides and toxic trace elements: prospects and limitations of *Paulownia tomentosa*. Toxics 10(8), Art. no 8 (2022). https://doi.org/10.3390/toxics10080465
58. Salman, M., et al.: Removal of cresol red by adsorption using wastepaper. Ind. Domest. Waste Manag. 2(1), Art. no 1 (2022). https://doi.org/10.53623/idwm.v2i1.63
59. Haleyur, N. et al.: Influence of bioaugmentation and biostimulation on PAH degradation in aged contaminated soils: Response and dynamics of the bacterial community. J. Environ. Manage. 238, 49–58 (2019). https://doi.org/10.1016/j.jenvman.2019.02.115
60. Abtahi, H., et al.: Effect of competition between petroleum-degrading bacteria and indigenous compost microorganisms on the efficiency of petroleum sludge bioremediation: field application of mineral-based culture in the composting process. J. Environ. Manage. 258, 110013 (2020). https://doi.org/10.1016/j.jenvman.2019.110013
61. Foong, S. Y., et al.: A recent global review of hazardous chlorpyrifos pesticide in fruit and vegetables: prevalence, remediation and actions needed, J. Hazard. Mater. 400, 123006 (2020). https://doi.org/10.1016/j.jhazmat.2020.123006
62. Chen, M., et al.: Study on soil physical structure after the bioremediation of Pb pollution using microbial-induced carbonate precipitation methodology. J. Hazard. Mater. 411, 125103 (2021). https://doi.org/10.1016/j.jhazmat.2021.125103
63. K. Cruz, J. Guézennec, et T. Barkay, Binding of Hg by bacterial extracellular polysaccharide: a possible role in Hg tolerance, Appl. Microbiol. Biotechnol., vol. 101, no 13, p. 5493-5503, juill. 2017, doi: https://doi.org/10.1007/s00253-017-8239-z.
64. Sravya, K., Sangeetha, S.: Feasibility study on bioremediation techniques to contaminated soils. Mater. Today Proc. 51, 2556–2560 (2022). https://doi.org/10.1016/j.matpr.2021.12.364
65. Soares, T. F. S. N., dos S. Dias, D. C. F., Oliveira, A. M. S., Ribeiro, D. M., dos S. Dias, L. A.: Exogenous brassinosteroids increase lead stress tolerance in seed germination and seedling growth of *Brassica juncea* L. Ecotoxicol. Environ. Saf. 193, 110296 (2020). https://doi.org/10.1016/j.ecoenv.2020.110296
66. Wang, M.-J., Sheng, G. D., Wang, R.: Progress in disposal technologies for plants polluted with heavy metals after phytoextraction. J. Agric. Resour. Environ. 38(2), 151–159 (2021). https://doi.org/10.13254/j.jare.2020.0756
67. Hassan, A., Pariatamby, A., Ossai, I. C., Hamid, F. S.: Bioaugmentation assisted mycoremediation of heavy metal and/metalloid landfill contaminated soil using consortia of filamentous fungi. Biochem. Eng. J. 157, 107550 (2020). https://doi.org/10.1016/j.bej.2020.107550

68. Sarma, H., Islam, N. F., Prasad, R., Prasad, M. N. V., Ma, L. Q., Rinklebe, J.: Enhancing phytoremediation of hazardous metal(loid)s using genome engineering CRISPR–Cas9 technology. J. Hazard. Mater. 414, 125493 (2021). https://doi.org/10.1016/j.jhazmat.2021.125493
69. Salehi, N., Azhdarpoor, A., Shirdarreh, M.: The effect of different levels of leachate on phytoremediation of pyrene-contaminated soil and simultaneous extraction of lead and cadmium. Chemosphere 246, 125845 (2020). https://doi.org/10.1016/j.chemosphere.2020.125845
70. Oladoye, P. O., Olowe, O. M., Asemoloye, M. D.: Phytoremediation technology and food security impacts of heavy metal contaminated soils: a review of literature. Chemosphere 288, 132555 (2022). https://doi.org/10.1016/j.chemosphere.2021.132555
71. Luo, J., Cai, L., Qi, S., Wu, J., Gu, X. W. S.: The interactive effects between chelator and electric fields on the leaching risk of metals and the phytoremediation efficiency of *Eucalyptus globulus*. J. Clean. Prod. 202, 830–837 (2018). https://doi.org/10.1016/j.jclepro.2018.08.130
72. Lei, Z., et al.: Biochar enhances the biotransformation of organic micropollutants (OMPs) in an anaerobic membrane bioreactor treating sewage. Water Res. 223, 118974 (2022). https://doi.org/10.1016/j.watres.2022.118974
73. Mousavi, A., Pourakbar, L., Siavash Moghaddam, S., Popović-Djordjević, J.: The effect of the exogenous application of EDTA and maleic acid on tolerance, phenolic compounds, and cadmium phytoremediation by okra (*Abelmoschus esculentus* L.) exposed to Cd stress. J. Environ. Chem. Eng. 9(4), 105456 (2021). https://doi.org/10.1016/j.jece.2021.105456
74. Kafle, A., Timilsina, A., Gautam, A., Adhikari, K., Bhattarai, A., Aryal, N.: Phytoremediation: Mechanisms, plant selection and enhancement by natural and synthetic agents. Environ. Adv. 8, 100203 (2022). https://doi.org/10.1016/j.envadv.2022.100203
75. Tipu, M. I. et al.: Growth and physiology of maize (*Zea mays* L.) in a nickel-contaminated soil and phytoremediation efficiency using EDTA. J. Plant Growth Regul. 40(2), 774–786 (2021). https://doi.org/10.1007/s00344-020-10132-1
76. Gunarathne, V. et al.: 7 - Phytoremediation for E-waste contaminated sites. In: Prasad, M. N. V., Vithanage, M., Borthakur, A., Éd., Handbook of electronic waste management, Butterworth-Heinemann, pp. 141–170 (2020). https://doi.org/10.1016/B978-0-12-817030-4.00005-X
77. Yadav, S. K., Pathak, R., Singh, R. K., Mahato, R. V.: Correlation of body mass index with waist circumference, random blood sugar and dietary pattern as predictors of diabetes mellitus. Int. J. Appl. Sci. Biotechnol. 6(3), Art. no 3 (2018). https://doi.org/10.3126/ijasbt.v6i3.21177
78. Ashraf, S., Ali, Q., Zahir, Z. A., Ashraf, S., Asghar, H. N.: Phytoremediation: environmentally sustainable way for reclamation of heavy metal polluted soils. Ecotoxicol. Environ. Saf. 174, 714–727, 2019. https://doi.org/10.1016/j.ecoenv.2019.02.068
79. Patra, D. K., Pradhan, C., Patra, H. K.: Toxic metal decontamination by phytoremediation approach: concept, challenges, opportunities and future perspectives. Environ. Technol. Innov. 18, 100672 (2020). https://doi.org/10.1016/j.eti.2020.100672
80. Hu, N., et al.: A novel method for determining the adequate dose of a chelating agent for phytoremediation of radionulides contaminated soils by *M. cordata*. J. Environ. Radioact. 227, 106468, (2021). https://doi.org/10.1016/j.jenvrad.2020.106468
81. Gavrilescu, M.: Enhancing phytoremediation of soils polluted with heavy metals. Curr. Opin. Biotechnol. 74, 21–31 (2022). https://doi.org/10.1016/j.copbio.2021.10.024
82. Vijayaraghavan, K., Reddy, D. H. K., Yun, Y.-S.: Improving the quality of runoff from green roofs through synergistic biosorption and phytoremediation techniques: a review. Sustain. Cities Soc. 46, 101381 (2019). https://doi.org/10.1016/j.scs.2018.12.009
83. Beecham, S., Razzaghmanesh, M.: Water quality and quantity investigation of green roofs in a dry climate. Water Res. 70, 370–384 (2015). https://doi.org/10.1016/j.watres.2014.12.015
84. Cascone, S.: Green roof design: state of the art on technology and materials. Sustainability 11(11), Art. No. 11 (2019). https://doi.org/10.3390/su11113020
85. Padmavathiamma, P. K., Li, L. Y.: Phytoremediation technology: hyper-accumulation metals in plants. Water Air Soil Pollut. 184(1), 105–126 (2007). https://doi.org/10.1007/s11270-007-9401-5

86. Ali, H., Khan, E., Sajad, M. A.: Phytoremediation of heavy metals—concepts and applications. Chemosphere 91(7), 869–881 (2013). https://doi.org/10.1016/j.chemosphere.2013.01.075
87. Sheoran, V., Sheoran, A. S., Poonia, P.: Phytomining: a review. Miner. Eng. 22(12), 1007–1019 (2009). https://doi.org/10.1016/j.mineng.2009.04.001
88. de Souza, T. D., Borges, A. C., Braga, A. F., Veloso, R. W., Teixeira de Matos, A.: Phytoremediation of arsenic-contaminated water by Lemna Valdiviana: an optimization study. Chemosphere 234, 402–408 (2019). https://doi.org/10.1016/j.chemosphere.2019.06.004
89. Kumar, U., Singh, R. S., Mandal, J., Nayak, A. K., Jha, A. K.: Removal of As(III) and Cr(VI) from aqueous solutions by *Bixa orellana* leaf biosorbent and As(III) removal using bacterial isolates from heavy metal contaminated site. J. Indian Chem. Soc. 99(5), 100334 (2022). https://doi.org/10.1016/j.jics.2021.100334
90. Hunt, A. J., et al.: Phytoextraction as a tool for green chemistry. Green Process. Synth. 3(1), 3–22, (2014). https://doi.org/10.1515/gps-2013-0103
91. Anderson, C. W. N., Bhatti, S. M., Gardea-Torresdey, J., Parsons, J.: In vivo effect of copper and silver on synthesis of gold nanoparticles inside living plants. ACS Sustain. Chem. Eng. 1(6), 640–648 (2013). https://doi.org/10.1021/sc400011s
92. Marques, A. P. G. C., Rangel, A. O. S. S., Castro, P. M. L.: Remediation of heavy metal contaminated soils: phytoremediation as a potentially promising clean-up technology. Crit. Rev. Environ. Sci. Technol. 39(8), 622–654 (2009). https://doi.org/10.1080/10643380701798272
93. Ginn, B. R., Szymanowski, J. S., Fein, J. B.: Metal and proton binding onto the roots of *Fescue rubra*. Chem. Geol. 253(3), 130–135 (2008). https://doi.org/10.1016/j.chemgeo.2008.05.001
94. Burges, A., Alkorta, I., Epelde, L., Garbisu, C.: From phytoremediation of soil contaminants to phytomanagement of ecosystem services in metal contaminated sites. Int. J. Phytoremediation 20(4), 384–397 (2018). https://doi.org/10.1080/15226514.2017.1365340
95. Göhre, V., Paszkowski, U.: Contribution of the arbuscular mycorrhizal symbiosis to heavy metal phytoremediation. Planta 223(6), 1115–1122 (2006). https://doi.org/10.1007/s00425-006-0225-0
96. LeDuc, D. L., Terry, N.: Phytoremediation of toxic trace elements in soil and water. J. Ind. Microbiol. Biotechnol. 32(11–12), 514–520 (2005). https://doi.org/10.1007/s10295-005-0227-0
97. Pilon-Smits, E. A., LeDuc, D. L.: Phytoremediation of selenium using transgenic plants. Curr. Opin. Biotechnol. 20(2), 207–212 (2009). https://doi.org/10.1016/j.copbio.2009.02.001
98. Guarino, F., Miranda, A., Castiglione, S., Cicatelli, A.: Arsenic phytovolatilization and epigenetic modifications in *Arundo donax* L. assisted by a PGPR consortium. Chemosphere 251, 126310 (2020). https://doi.org/10.1016/j.chemosphere.2020.126310
99. Kumar, B., Smita, K., Cumbal Flores, L.: Plant mediated detoxification of mercury and lead. Arab. J. Chem. 10, S2335–S2342 (2017). https://doi.org/10.1016/j.arabjc.2013.08.010
100. Muthusaravanan, S., et al.: Phytoremediation of heavy metals: mechanisms, methods and enhancements. Environ. Chem. Lett. 16(4), 1339–1359 (2018). https://doi.org/10.1007/s10311-018-0762-3
101. Singh, S., Santal, A. R.: Phytoremediation of heavy metals: the use of green approaches to clean the environment. Phytoremed. Manag. Environ. Contam. 2, 115–129 (2015)
102. Wuana, R. A., Okieimen, F. E.: Heavy metals in contaminated soils: a review of sources. Chemistry, risks and best available strategies for remediation. Int. Sch. Res. Not. 2011, e402647 (2011). https://doi.org/10.5402/2011/402647
103. Islam, Md. S., Ueno, Y., Sikder, Md. T., Kurasaki, M.: Phytofiltration of arsenic and cadmium from the water environment using *Micranthemum umbrosum* (J.F. Gmel) S.F. Blake As A Hyperaccumulator. Int. J. Phytoremediation 15(10), 1010–1021 (2013). https://doi.org/10.1080/15226514.2012.751356
104. Fine, P., Rathod, P. H., Beriozkin, A., Mingelgrin, U.: Uptake of cadmium by hydroponically grown, mature *Eucalyptus camaldulensis* saplings and the effect of organic ligands. Int. J. Phytoremediation 15(6), 585–601 (2013). https://doi.org/10.1080/15226514.2012.723061

105. Rana, V., Maiti, S. K.: Municipal and industrial wastewater treatment using constructed wetlands, in phytoremediation: in-situ applications. In: Shmaefsky, B. R., Éd., Concepts and strategies in plant sciences. Cham: Springer International Publishing (2020), pp. 329–367. https://doi.org/10.1007/978-3-030-00099-8_10
106. Brewer, C. E., et al.: New approaches to measuring biochar density and porosity. Biomass Bioenergy 66, 176–185, (2014). https://doi.org/10.1016/j.biombioe.2014.03.059
107. Ozyigit, I. I., Can, H., Dogan, I.: Phytoremediation using genetically engineered plants to remove metals: a review. Environ. Chem. Lett. 19(1), 669–698 (2021). https://doi.org/10.1007/s10311-020-01095-6

Advanced Technologies for Transforming Biomass to Biofuels

Shehbaz Ali, Abida Rani, Sivasamy Sethupathy, Fakhra Liaqat, Wang Shunkai, Tawaf Ali Shah, and Daochen Zhu

Abstract This chapter explores biomass and biofuels as renewable energy sources. Before biofuels replace fossil fuels, food security, land use, and environmental impact must be addressed. To maintain ethical and ecological production and consumption, biomass and biofuel sources and technologies must be carefully evaluated. Fourth-generation biofuels, which are produced through the genetic engineering of microorganisms, have been found to exhibit superior qualities compared to the previously discussed generations. This chapter also explores a range of potential sources of bioenergy, encompassing agricultural and forestry residues, energy crops, algae, municipal solid waste (MSW), and animal waste. The cultivation of algae, anaerobic digestion, combustion, gasification, and pyrolysis, as well as their associated benefits and drawbacks, are also addressed in this chapter. The utilization of bioenergy holds promising potential in mitigating environmental impact while simultaneously stimulating economic growth. It is imperative to ensure that these processes are conducted in a sustainable manner, without causing any adverse effects on agricultural output or other land uses. Hence, future research should focus on improving renewable energy production through sustainable biomass production.

Keywords Biofuel generations · Anaerobic digestion · Algae biomass

S. Ali (✉) · S. Sethupathy · F. Liaqat · W. Shunkai · D. Zhu
Biofuels Institute, School of the Environment and Safety Engineering, Jiangsu University, Zhenjiang 212013, Jiangsu, China
e-mail: shehbaz205@googlemail.com

A. Rani
Department of Pharmaceutical Chemistry, Faculty of Pharmacy, Bahauddin Zakariya University, Multan, Pakistan

T. A. Shah
College of Agriculture Engineering and Food Science, Shandong University of Technology, Zibo, China

J. Mabrouki and M. Azrour (eds.), *Integrated Solutions for Smart and Sustainable Environmental Conservation*, Studies in Systems, Decision and Control 527,
https://doi.org/10.1007/978-3-031-55787-3_4

1 Introduction

Modern societies value energy reliability. Feedstock production involves growing biomass from plant, animal, or microbial cells to use as fuel. Sugarcane, soybeans, maize, sugar beets, proteins, fats, starches, lignocellulosic constituents, agricultural waste, and forest residues are biomass that can be used as a feedstock. Biomass to bioenergy conversion process produces first- and second-generation biofuels of these biomasses. Due to their lower greenhouse gas emissions, biofuels outperform fossil fuels, hence can replace fossil fuels [1, 2]. Biofuel production is heavily influenced by biomass composition. Despite its many environmentally sustainable applications, this technology could worsen food insecurity, land use changes, water scarcity, energy consumption, and deforestation. Characterizing biomass before using it is essential. The current goal is to analyses bio-energy processes and carefully weigh their pros and cons. According to studies this method ensures ethical and sustainable production [2, 3].

The current areas of investigation revolve around third- and fourth-generation biofuels. Various species of macro and microalgae have the ability to utilize sunlight and carbon dioxide in order to synthesize biomass. The aforementioned procedure has the potential to be employed in the production of third-generation biofuels and other valuable commodities [3–5]. Synthetic biology is an academic discipline that encompasses the deliberate manipulation of genetic material and metabolic pathways within microorganisms, with the aim of generating biofuels classified as fourth-generation. The implementation of meticulous engineering practices has facilitated the production of fourth-generation biofuels that adhere to the specifications outlined by various fuel classifications. Enhancing the cost-effectiveness and efficiency of fourth-generation biofuels can be achieved through the utilization of a broader range of feedstocks and the mitigation of production expenses. Several benefits have been mentioned above, including increased food availability, efficient land utilization, and enhanced sustainability [3].

This chapter is to evaluate a range of biomass and biofuel sources and technologies with the aim of promoting and enhancing sustainable production and utilization. Figure 1 is a schematic of biomass resources that shows how biomass can be transformed into bioenergy through different processes and what each process is used for.

2 Biomasses Resources

Sources of renewable biomass energy include plants, energy crops, municipal waste, wet waste, and algal byproducts. Utilizing waste from landfills and garbage dumps has the potential to produce renewable energy. As reliable sources of energy, corn kernels and stalks, soybean and canola oils, animal fats, prairie grasses, hardwoods, and algae are all recognized. A variety of energy products, including electricity,

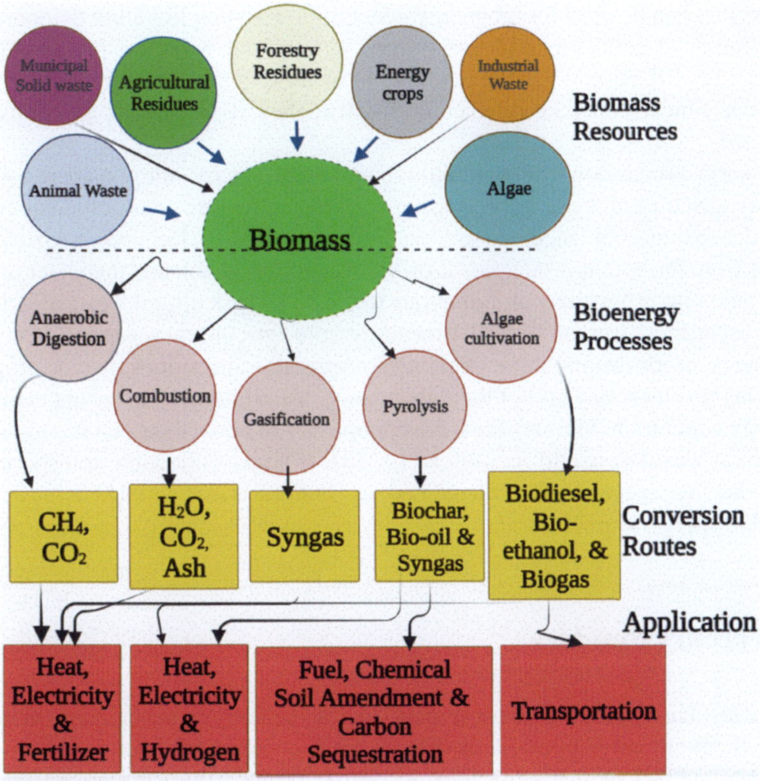

Fig. 1 A basic schematic representation of biomass resources, the conversion of biomass into bioenergy through various processes and their applications

liquid and solid fuels, gaseous fuels, heat, chemicals, and other energy commodities are produced using biomass energy technologies by utilizing renewable organic resources. The primary renewable energy source on a global scale is biomass. Bioenergy accounts for approximately 55% of all renewable energy sources [6–8].

2.1 Agricultural Residues

Agricultural waste is abundant and easy to access, making it a practical alternative for use as a biomass source for the production of bioenergy. Feedstock for the production of bioenergy and biofuels can be derived from a wide variety of sources, including crop wastes, industrial byproducts, and animal excrement production [9]. Feedstock for bioenergy production often consists of crop wastes like maize stover, wheat straw, and rice straw. Post-harvest agricultural leftovers left behind. Bagasse from sugarcane milling and rice bran and hulls from rice milling are two examples of processing leftovers that can be used for bioenergy generation. Existing literature demonstrates the feasibility of collecting and transporting agricultural waste for processing into bioenergy [9, 10]. Biogas has the potential to be used in the generation of both heat and power which reduce on reliance on fossil fuels and greenhouse gas emissions [11].

Bioenergy production from agricultural waste requires specific considerations. To transport agricultural waste to processing facilities in remote areas, efficient collection and transportation methods are needed economically. Removing agricultural residues from the soil may decrease crop yields and soil health and then later need to add organic matter/fertilizer and minerals to the soil. Agricultural waste affects soil quality, according to research [5, 7]. Bioenergy from agricultural waste can meet long-term energy needs despite these challenges. Agricultural residues include crop and processing residues. Because of their abundance, these alternatives are appealing for bioenergy generation. Farmers have the opportunity to boost their income by selling agricultural waste as useful feedstock [12, 13]. If better collection, transportation, and residue management are implemented, agricultural waste could become a major renewable energy source.

2.2 Forestry Residues

Forestry residues including sawdust, bark, and branches are abundant and versatile, making them attractive biomass feedstocks for bioenergy production. Due to their abundance and lack of alternative uses, forestry byproducts have drawn considerable interest as a possible biomass source for bioenergy. By removing superfluous materials from the forest floor, the use of forestry residues for bioenergy helps reduce the risk of forest fires and speeds up the adoption of sustainable forest management techniques [13]. Their dispersed and often inaccessible locations, finding and transporting forest remains can be prohibitively expensive. Specialized vehicles and portable chippers are only two examples of the technological advancements that have made the collection and delivery of leftover forest debris more efficient and less expensive [13, 14].

There are a lot of things to think about before transforming forest waste into bioenergy such as think about how forest residues are categorized, their quality and chemical composition, the costs of harvesting and processing, the effects on the environment and the market, and any technological developments. High-moisture residues may require drying or other processing before being put to use. Potentially unusable for fuel production are residues with high levels of contaminants. Its lignin and ash contents determine the suitability of residues for bioenergy processes [15]. Using these criteria, forest remnants that are well suited for bioenergy production can be located and environmentally sound management techniques implemented. Reducing greenhouse gas emissions and increasing sustainable energy production are both possible when forest wastes are used as biomass feedstock for bioenergy production [16].

2.3 Energy Crops

Energy crops are cultivated specifically for biomass production. These plants are utilized in the production of first-generation fuels. The selection of energy crops for biomass is determined by factors such as their rapid growth rate, per capita yield, and compatibility with soil attributes conducive to crop growth. The cultivation of energy crops serves an additional purpose, which includes the promotion of environmentally-friendly farming practices and providing indirect financial support to farmers [6, 16]. Sugar cane, sugar beetroot, corn, sweet sorghum, oats, barley, rye, rapeseed oil, and soybeans are widely recognized as prominent energy crops. However, it is worth noting that switchgrass and miscanthus are the most prevalent among them [3, 16]. This phenomenon can be attributed to the rapid growth rate, effective utilization of resources, and consistent reproductive capabilities. In addition to the widely recognized non-woody crops used as energy sources, it is worth noting that woody plants such as poplar, willow, and eucalyptus possess similar characteristics to grasses, including rapid growth and cultivation [6, 17]. Therefore, they have been utilized as renewable energy sources. A common example of energy crops is; utilization of sugarcane and sweet sorghum as feedstocks for the production of ethanol. Previous research has been shown that these specific plant species demonstrate strong growth in tropical and subtropical climates [17].

There are many factors that affect biomass production of an energy crops. When making agricultural decisions, it is essential to consider crop diversity and quality, expected yield, environmental impacts, and financial profitability. Water and land use for energy crops must be carefully managed due to the high demand for these crops [18]. Despite these challenges, the potential for technological advancement lies in the utilization of biomass production from energy crops. Concerns about food insecurity may arise when the production of food and feed is hampered, as an example, by the cultivation of energy crops. Soil nitrogen loss and an increase in the use of synthetic fertilizers and pesticides are two examples of the unintended negative effects on the environment that could result from the development of energy crops pesticides [19]. Thus, energy crops could quickly become a bioenergy industry resource. This would help farmers financially and promote environmentally responsible land management and greenhouse gas emissions reduction. Energy crop sustainability depends on crop diversity, quality, yield potential, environmental impacts, and economic viability. Energy crops must overcome several production challenges before they can replace fossil fuels as long-term, sustainable, and profitable bioenergy sources.

2.4 Algae

Biomass from both macro- and microalgae can be converted into biodiesel and biofuels. This sustainable biomass option can be stored for a long time. The lipid content of rapidly growing microalgae is quite considerable [20]. Biodiesel is produced from lipids by transesterification. Microalgae can be grown in seawater, wastewater, and non-arable soil and can be grow in photobioreactors; then processed into biodiesel. Biofuels can be processed from seaweed or macroalgae. Carbohydrate-rich macroalgae can be fermented into biofuels. In contrast to most biofuel crops, macroalgae can really be grown in saline water [20, 21]. Moreover, microalgae and macroalgae are superior to traditional biofuel crops in terms of production, growth rate, and tolerance of non-arable soil and wastewater. Due to its ability to both supply nutrients and trap carbon, algae farming is often compatible with other land uses [22]. However, algae-based biofuels have not yet been commercialized because of their high production costs and a lack of research and development. Despite these obstacles, biodiesel production from macro and microalgae remains feasible. More research is needed, but they show promise as a long-term, environmentally benign power options [23].

Algal biomass can be subjected to various processing methods, such as pyrolysis, hydrothermal liquefaction, or anaerobic digestion, depending on the appropriate conditions [3]. The superiority of this kind of innovation lies in its ability to utilize saline water and unproductive land, thereby minimizing competition with food crops. However, the cultivation of this faces certain challenges, such as water scarcity and interference from other organisms, rendering its growth unfeasible in arid environments. The growth of plants can be impeded by temperature and climate as well. The production of bioenergy necessitates the cultivation of biomass in an economically viable manner. The potential of bioenergy necessitates additional investigation [3, 24]. The meticulous examination and investigation of the enduring ecological consequences resulting from human activities is imperative.

2.5 Municipal Solid Waste (MSW)

MSW includes food, yard, and household waste. Anaerobic digestion, combustion, and gasification are the most common bioenergy processes used on the organic material in municipal solid waste. The final product of these processes is a mixture that looks like other feedstocks. Some MSW requires supplementary aid for sorting organic and inorganic components to make it suitable for bioenergy process. Thus, prior to using this as a feedstock, it is crucial to implement suitable classification and management protocols. The biogas or syngas that can be produced from these wastes has the same applications and benefits as the conventional fossil fuels that they replace [4, 25].

There are numerous advantages to producing municipal solid waste bioenergy and biofuels. For example, these waste conversion to bioenergy can reduce landfill waste, greenhouse gas emissions, and the use of fossil fuels [4, 20]. Poor waste management could increase the cost of renewable energy by making it less efficient at converting MSW to usable forms [26]. The conversion process and system economics must be thoroughly examined to ensure that the benefits outweigh the costs and difficulties.

2.6 Animal Waste

Animal manure, fat, and other byproducts are valuable feedstock options for the production of bioenergy. The utilization of these wastes as a feedstock is increasing attention due to its ability to address two significant societal concerns: the provision of energy and the mitigation of environmental contamination through waste management. However, it is noteworthy that this process yields similar types of bio-energy products and offers comparable benefits to other waste management methods. These benefits include reduced waste generation, increased production of renewable energy, and decreased emissions of greenhouse gases [20, 27].

These waste materials can undergo various energy processes. However, they are particularly valuable for biodiesel production due to their high nitrogen, phosphorus, potassium, and fat residue contents. They can be utilized as a diesel substitute without any modifications [27]. In addition, the production of syngas is a notable outcome resulting from the utilization of these waste materials. Like other feedstocks, it also possesses significant potential for generating renewable energy that can serve as a viable alternative to fossil fuels, while simultaneously contributing to waste reduction efforts. In addition to the aforementioned benefits, it provides financial support to individuals residing in rural areas [11, 27].

However, these waste materials also have certain disadvantages. For instance, a lower carbon-to-nitrogen ratio in their chemical composition can impede the process by increasing alkalinity, rendering them unsuitable for microorganism growth and digestion. Their composition is contingent upon factors such as the animals' feeding habits and the prevailing weather conditions. The expenses associated with transportation, inadequate management practices, and liquid manure handling can contribute to water pollution. Additionally, the use of manure may compete with soil amendment substrates [11, 27]. The practicality of bioenergy production is evident due to its potential to enhance sustainable agricultural practices, promote energy production, and effectively mitigate environmental impacts. When deciding on biomass selection, it is imperative to consider factors such as availability, price, and environmental impacts.

3 Biomasses Resources

Bioenergy would be capable of reducing greenhouse gas emissions and thus mitigating the effects of climate change. This can help community's growth by increasing employment opportunities and hastening rural expansion. However, these resources must be produced in a way that is both sustainable and does not have a negative impact on food production or other land usage. It is critical to recognize that different bioenergy processes have varying effects on natural and human communities. Monocultures of energy crops may have negative effects on soil quality, water supplies, and biodiversity. There is also the possibility of agricultural land being relocated and wildlife habitats being destroyed. Additional carbon emissions may be produced by long-distance feedstock transportation and the high energy requirements of some bioenergy processes. The study's primary focus could be on the environmental and social costs of biomasses and energy conversion methods, as well as their effects on the local population and the economy as a whole. Let us engage in a comprehensive examination of various bioenergy processes, encompassing both biochemical processes such as anaerobic digestion, as well as thermodynamic processes including pyrolysis, gasification, and combustion.

3.1 Anaerobic Digestion (AD)

Biogas is produced when microorganisms digest organic matter in the absence of oxygen. Aqueous biomass and bacteria are used in an enclosed digester for AD to take place. It can be produced from a variety of sources, including manure, food waste, sewage sludge, and agricultural residues [27]. Once AD is complete, the digestate can be used as a nutrient-rich plant fertilizer. This strategy reduces the need for environmentally harmful artificial fertilizers. However, AD is not without its limitations and this process needs to be carefully managed to ensure the highest possible digester efficiency and nutrient balance in the feedstock [11, 27]. Typically, there are four phases: Hydrolysis (phase 1), the process by which complex organic compounds are broken down into amino acids, sugars, and fatty acids, is sped up by microorganisms that produce enzymes. In the process of acidogenesis (phase 2), microorganisms aid in the transformation of simpler molecules into organic acids. Acetogens (Phase 3) are involved the conversion of organic acids by microorganisms into acetic acid, hydrogen, and carbon dioxide. During methanogenesis (phase 4), methanogenic bacteria convert carbon dioxide and hydrogen into methane [11, 27].

Biogas from AD has advantages such as heating buildings, electricity generation, vehicles fuel production and fertilizer. This process reduces the waste disposal costs and does not emit no greenhouse gases. Unlike conventional energy sources, renewables can be replenished indefinitely and do not contribute to pollution. Less money is spent on garbage pickup [7, 27, 28]. Some problems with the AD process are that it needs constant feedstock supply, controlled process temperature, balanced carbon to nitrogen ratio. Hence, this technology is expensive to set up and run. Even with these problems, anaerobic digestion could help the world get more energy [27].

3.2 Combustion

Combustion is a thermodynamic process wherein biomass is oxidized to generate thermal energy (in the form of heat and steam) or electrical energy. This process is widely acknowledged and embraced in many countries. The process requires a temperature range of 700–1200 °C in the presence of oxygen. A range of feedstocks that have the potential to undergo combustion, encompassing wood, agricultural waste, and municipal solid waste. The resultant products and their respective applications exhibit a level of comparability akin to that observed in other processes. This process comprises three discrete phases in order to attain full combustion. The first stage, commonly known as phase 1, involves the drying/dehydration of the biomass. During the second phase, referred to as phase 2, an additional amount of heat is necessary to facilitate the vaporization of the feedstock and its subsequent conversion into gaseous components. In the final phase (phase 3) of the experiment, the production of heat occurred as a result of the combustion of gaseous chemicals in the presence of oxygen [9, 29].

However, this process has several drawbacks. It releases nitrogen, carbon monoxide, and dioxide, polluting the air. This process also pollutes water with ash. Transporting and managing this material can also be difficult. In liquid feedstocks that require drying, the process under consideration is less efficient [29, 30]. Biomass combustion research should focus on improving combustion efficiency, which can reduce emissions. Innovative combustion technology and careful parameter optimization can reduce biomass burning pollutant emissions.

3.3 Gasification

Gasification is an emerging technological approach that facilitates the conversion of biomass into bioenergy through the utilization of thermochemical reactions. The process involves exposing the biomass to temperatures ranging from 500 to 1400 °C, pressures of up to 33 bars, and anaerobically, or with little oxygen environment. The final outcome of this procedure yields syngas, a composite of diverse gases such as carbon monoxide, hydrogen, carbon dioxide, and methane [25, 31]. Agricultural residues, wood, and waste materials are just some of the biomasses that can be converted by gasification [3]. This process can achieve a relatively efficient conversion efficiency of up to 80%. There are three main stages in this process. The first step is to dry the biomass to a moisture content of 20% or less for smooth functioning of process. Pyrolysis is a second stage that refers to the oxygen-free thermal decomposition of biomass [11, 31]. Pyrolysis gas and char are the two main byproducts of this process. Gasification is the third and final stage of the process, and it involves the reaction of the pyrolysis gas with steam or oxygen. Syngas and very little char are the byproducts of the aforementioned process [9, 25].

It can be carried out in a wide range of reactors, such as fixed-bed reactors, fluidized-bed reactors, and entrained-flow reactors. The input material and the final product determine the type of reactor used in biomass processing. The syngas produced can be used to power vehicles or generate electricity [25, 31]. One of the many advantages of gasification as a renewable energy source is its high efficiency. The equipment is capable of efficiently managing a variety of biomass sources. However, the implementation of this technology faces multiple challenges, such as the high cost of acquiring the necessary equipment, the production of harmful byproducts, and the relatively recent emergence of other technologies in comparison to other bioenergy solutions. Opportunities exist in the fields of pollutant emission regulation, the exploration of innovative syngas applications, and the development of economically viable technologies [25, 31]. This technology has both positive and negative characteristics, and it can be more widely adopted if its challenges are addressed.

3.4 Pyrolysis

Pyrolysis, a thermochemical process that takes place in the absence of oxygen, entails the thermal degradation of biomass into three distinct phases: solid, liquid, and gaseous products. This process can be conducted effectively within a temperature range spanning from 350 to 700 °C. This approach yields three primary products: char, which is the solid component; bio-oil, which is the liquid component; and syngas, which is the gaseous component [32]. It has the potential to effectively utilize a wide range of feedstock, such as animal waste, (MSW), wood, and agricultural residues [3, 32]. The technology for converting biomass into biofuels has a conversion efficiency of 70%. It can be categorized into three distinct phases. During the initial stage, the moisture content of biomass is decreased to a level below 20% to ensure that it does not hinder the progress of the process. The second stage involves the oxygen-free thermal decomposition of biomass. Two major byproducts of pyrolysis are char and pyrolysis gas. Condensation is the last step in the pyrolysis process, in which the bio-oil and syngas are extracted [21, 32].

For the pyrolysis process, various reactor types including fixed-bed, fluidized-bed, and entrained-flow reactors, are available. When selecting a reactor, the type of biomass and the desired product must be considered. These items are applicable in numerous contexts, including commercial and residential construction. Among these applications are automobiles, fuel cells, and the generation of electricity [3]. It has advantages over combustion is cost-effectiveness and its limited pollutant emissions [3, 21]. As a result, it has been proposed that future research concentrate on identifying more refine and cost-effective methods of pollution control.

3.5 Algae Cultivation

Algae are incredibly diverse and widespread as photosynthetic organisms in both aquatic and terrestrial ecosystems. These sources are promising for bioenergy production due to their high biomass yield and scalability. Open ponds or closed photobioreactors can be used for algae cultivation. Open cultivation methods, in which the algae are grown in ponds or tanks filled with water, are used by the vast majority of algae farms today. This method is less effective because it subjects algae to environmental stressors like sunlight and wind, both of which can stunt their growth

[20, 22]. A photobioreactor is a hermetically sealed container used for growing plants in a special growth medium. This method is preferred because it protects the algae from sunlight and wind, two factors that can stunt their development [20].

Algae can produce up to 100 g/m^2 of biomass [22, 23]. Biodiesel, bioethanol, and biogas are all products that could be left over from producing algal biomass. High cultivation costs, low conversion efficiency, and the possibility of eutrophication are just a few of the difficulties that arise when growing algae for bioenergy production [20]. Research pertaining to algal cultivation for bioenergy production exhibits promising results. Although advancements have been achieved in the reduction of cultivation expenses and the enhancement of conversion efficiency, further efforts are required to mitigate the environmental consequences associated with this practice.

4 Generation of Biofuels

Biomass and biofuels have come under intense scrutiny in recent years as possible replacement for depleting and climate-altering fossil fuels. Their production and use raise concerns about their possible influence on food security, land use, and the environment, despite their many advantages, including reduced greenhouse gas emissions and air pollution. Consequently, it is essential to thoroughly weigh the pros and cons of various biomass and biofuel sources and technologies to guarantee their moral and ecological production and usage in the transition to a more sustainable energy future. Here, discuss the generation of biofuel and suitability of biomass as well as the difficulties and potential gains associated with putting these technologies into practice. Table 1 presents an overview of a source discussing the application of bioenergy and the process involved in generating biofuels.

4.1 First Generation of Biofuels

First-generation biofuels refer to the liquid fuels made from cereal grains, oilseeds, and sugar beets. There are number of feedstocks can be used to produce it such as sugars, starches, and oils. Its main advantages include the production of renewable energy and its incorporation into energy infrastructure and transportation systems [5]. As we know that it is ecofriendly process but it causes very serious concern related food security [5, 33]. Recently a debate on this generation is already started due to its competition among food production industry and biofuel process may cause inflation of food crops.

However, besides food security, there is one example of environmental pollution by this generation is producing nitrous oxide greenhouse gases during ethanol production of from maize [5, 34]. And also, there is more expenditure in terms of energy, land and time in first generation biofuels, that's why other three generation are more efficient by using non-food crops and algae [5, 33]. First-generation

Table 1 A comprehensive overview of generation biofuels

Generation	Source	Bioenergy process	Main product	Applications	Remarks
First	Oil seed (Soybeans, Rapeseed, Palm oil), Cereal grains (wheat) Sugarcane, Corn	Biochemical process; Fermentation, Enzymatic hydrolysis, Transesterification. Thermodynamic methods; Pyrolysis	Bioethanol and Biodiesel	Vehicles fuel	It raises the issue of food versus fuel
Second	Non-edible part of food, Lignocellulosic biomass (wood, agricultural & forestry residues,)	Biochemical process; Anaerobic digestion, Thermodynamic process; Pyrolysis, Gasification, combustion	Bioethanol, Biodiesel, Fischer–Tropsch fuels, Syngas	Vehicles fuel, Power generation, Heating, steam, Electricity	Converting biomass into liquids and gases is a clean and renewable energy source
Third	Biomass from algae and microorganisms can be grown quickly and efficiently without the use of arable land	Biochemical process; Anaerobic digestion, Thermodynamic process; Pyrolysis, Gasification, combustion	Bio-oil, Bioethanol syngas, hydrogen,	Vehicles fuel, Power generation, Heating, steam, Electricity	It does not require land or fertilizer, pesticides but it does necessitate a downstream process
Fourth	The genetic manipulation of algae and microorganisms to enhance biomass growth through engineering techniques	Biochemical process; Anaerobic digestion, Thermodynamic process; Pyrolysis, Gasification, combustion	Bio-oil, Bioethanol syngas, hydrogen,	Vehicles fuel, Power generation, Heating, steam, Electricity	It also does not require land, fertilizer, or pesticides. Investment in resources and the use of genetic engineering methods are necessary to improve carbon dioxide capture efficiency and boost crop yield. Downstream processing issues are dealt with

biofuels boosted bioenergy growth despite their flaws. Renewable energy has helped to reduce fossil fuel dependence and accelerate climate change adaptation. Biofuels have potential as a long-term energy source, but they need more lab research [33, 34].

4.2 Second-Generation Biofuels

The production of second-generation biofuels involves the utilization of energy crops, agricultural and forestry residues, as well as waste. Switchgrass, miscanthus, lignocellulosic biomasses, and wood are among the various feedstock options available for utilization [5, 16]. The production of second-generation biofuels involves the utilization of more intricate plant components, including cellulose, hemicellulose, and lignin [5, 35]. This technology possesses the potential to mitigate our dependence on fossil fuels owing to its favorable impact on efficiency, pollution reduction, and food production. The production of second-generation biofuels involves gasification, pyrolysis, and AD. The generation of ethanol, biodiesel, biogas, and other biofuels is the ultimate outcome of this [16, 33, 35].

As biofuel production does not compete with food production, it has the potential to reduce food insecurity. Thus, this generation has a lower environmental impact because it does not use pesticides or fertilizers. It also helps to preserve land and water. These products have no negative effects on the environment or land use [33, 35]. They are distinct from first-generation biofuels. Bioenergy has the potential for operational and environmental growth. Because of their higher energy density, second-generation biofuels produce more power per unit volume [33]. Hydrocarbons and sugars can be extracted from complex plant materials using modern extraction methods. Commercialization is difficult due to resource constraints, and affordability necessitates perseverance.

4.3 Third-Generation Biofuels

Algae and other microorganisms offer a viable solution to the issues of associate with previous generations of biofuels by producing biomass with a high lipid content that may be transformed into biodiesel and other fuels [36]. Microorganisms used in the production of third-generation biofuels do not require freshwater for growth, unlike crops used in the production of two earlier discussed biofuels [5, 33, 36, 37]. The extracted algae oil can be used to produce biodiesel-like qualities by transesterification and other biomass can be converted to biogas by AD [5, 33]. The use of algal feedstock has potential advantages like less environmental effects; no need for specific fertilizers or pesticides; and no interaction with food production [36]. After lipid extraction, the remaining waste provides an opportunity to create valuable byproducts, such as protein-rich animal feed and nutrient-rich fertilizer, from third-generation biofuels. The use of this tactic could cut down on the money spent on landfills and open up new markets for biofuel producers [36, 37].

Several challenges must be addressed by the present generation of biofuel technology before it can be deemed genuinely efficient. To achieve cost-effective fuel production, it is imperative to surmount obstacles such as the exorbitant expenses associated with the developmental phase of lipid extraction and refining technology. Controlling environmental factors to prevent fluctuations in algae and microbial productivity is another prerequisite for commercialization [36, 37]. Through consistent technological advancements, there is potential for a promising future for this alternative energy source, which could serve as a viable substitute for traditional fossil fuels in the near future.

4.4 Fourth-Generation of Biofuels

Synthetic biology-based biofuels are the fourth-generation. Genetic engineering and synthetic biology generate microorganisms that can produce specific biofuels or fuel precursors. Since they can be tuned to specific fuel types and manufactured utilizing a variety of feedstocks, fourth-generation biofuels may be more efficient and cost-effective than earlier generations. However, fourth-generation biofuels are still in under development, and synthetic biology-based technologies may have environmental and social implications [5]. Synthetic biology can engineer the enhancement of CO_2 fixation, lipid production, and extremophilic in living organisms [5, 38]. Renewable and environmentally-friendly, these fuels score well on the global warming potential scale [39]. The fuels and gases produced have low emissions and can be used sustainably. In the fourth generation of biofuels, modified microorganisms are also able to use in conjunction with genetically modified feedstock. Biofuel production from cyanobacteria has been boosted through genetic engineering [5, 38, 39]. Similarly, the quality and yield of ethanol-producing organisms are improved through genetic engineering. Producing biofuels from third- and fourth-generation feedstocks appears feasible and sustainable. To properly evaluate the potential of the topic, more studies are required [28, 39]. However, Current biofuels require efficiency optimization, cost-effective production methods, genetically engineered microorganism stability and safety, regulatory approval, and environmental and health impact assessments. These issues can be solved through collaboration between active researchers, industry partners, and regulatory agencies.

Third- and fourth-generation biofuels research has potential to help create a more sustainable energy system. Algae and synthetic biology-based technologies are more efficient and cost-effective than second-generation biofuels and provide environmental benefits. To ensure sustainable and responsible biofuel production and use, it is crucial to assess the pros and cons of various biomass and biofuel sources and technologies. Food security, land use change, and environmental impacts are the challenges [5, 27, 40].

5 Recent Advancements

The field of nanotechnology research has significant potential for improving the efficiency and cost-effectiveness of the biomass-to-biofuel conversion process. Nanoparticles with dimensions less than 100 nm show promising potential as catalysts in a variety of chemical reactions. This is due primarily to their ability to increase the surface area of biomass, facilitating enzymatic hydrolysis, and to reduce the viscosity of bio-oil. Improving the efficacy and efficiency of the biomass-to-biofuel conversion process holds promise for reducing the burden on global energy and material resources [40–42]. BASF, a German multinational chemical company, has been at the forefront of using nanotechnology and currently investigating the feasibility of using nanostructured catalysts to improve the efficiency of biomass conversion processes. BASF is currently also working on novel methodologies to improve the efficiency of lignocellulosic biomass-derived biofuel production. Oil refineries have prioritized the strategic goal of diversifying their energy sources in their portfolios by allocating resources to the development and implementation of renewable energy technologies. The aforementioned companies are currently allocating resources to R&D initiatives aimed at improving the ecological integrity, efficacy, and cost-effectiveness of gasification techniques [40, 42]. Researchers have also developed advanced technologies for the direct conversion of lignocellulosic biomass into sugar, bypassing enzymatic and chemical processes [43]. Novozymes, a Danish biotechnology company, is currently developed highly efficient enzyme in the field of lignocellulosic degradation [44].

The exploration of alternative approaches for converting biomass into biofuels has the potential to substantially diminish our reliance on fossil fuels. The conversion of biomass into biofuels is anticipated to result in enhanced efficiency and increased economic feasibility. The implementation of these techniques is expected to result in the development of more efficient and environmentally sustainable biofuel alternatives. Biobutanol, biohydrogen, and syngas exhibit considerable potential as biofuel alternatives.

6 Conclusion and Future Perspectives

Biofuels could replace fossil fuels and protect the environment. First-generation biofuels have drawbacks such as deforestation and greenhouse gas emissions. Even though third and fourth-generation biofuels may be more cost-effective and efficient than fossil fuels, second-generation biofuels are still considered environmentally friendly. These technologies must be carefully and thoroughly developed to succeed and not harm to food safety, land use, or the environment. As we move towards a sustainable energy model, we must carefully evaluate biomass and biofuel sources and technologies. This evaluation ensures these resources are created and used ethically and sustainably. To combat climate change and reduce the use of nonrenewable resources, renewable energy must be prioritized. This has led to exciting bioenergy and feedstock advances! Biofuels can expand! This can be achieved by focusing on advanced biofuel technology research and development, exploring different feedstock, working with other industries, and encouraging more people to use biofuels for transportation and energy can achieve this. The global bioenergy market is expected to reach $250 billion by 2025. Bioenergy will supply 14% of the global primary energy demand by 2050, which is great! To achieve sustainable development, we must recognize and address many challenges. We must address financial viability, environmental impact, and competition for land and water resources. Bioenergy production can be made more efficient and sustainable by using waste as a feedstock, combining bioenergy with other renewable energy sources, and using new technology. Bioenergy storage technologies, though still developing, may help manage energy delivery fluctuations. Bioenergy has great potential, but some challenges must be overcome before it can be used everywhere.

Acknowledgements The author is thankful to Prof. Daochen Zhu for providing postdoctoral fellowship, guidance, and support in the writing of this book chapter.

References

1. Srivastava, R.K., Shetti, N.P., Reddy, K.R., Kwon, E.E., Nadagouda, M.N., Aminabhavi, T.M.: Biomass utilization and production of biofuels from carbon neutral materials. Environ. Pollut. **276**, 116731 (2021)
2. Rehman, Z.U., Anal, A.K.: Enhanced lipid and starch productivity of microalga (Chlorococcum sp. TISTR 8583) with nitrogen limitation following effective pretreatments for biofuel production. Biotechnol. Rep. **21**, e00298 (2019)
3. Ambaye, T.G., Vaccari, M., Bonilla-Petriciolet, A., Prasad, S., van Hullebusch, E.D., Rtimi, S.: Emerging technologies for biofuel production: A critical review on recent progress, challenges and perspectives. J. Environ. Manage. **290**, 112627 (2021)
4. Kiehbadroudinezhad, M., Merabet, A., Ghenai, C., Abo-Khalil, A.G., Salameh, T.: The role of biofuels for sustainable MicrogridsF: A path towards carbon neutrality and the green economy. Heliyon (2023)
5. Alalwan, H.A., Alminshid, A.H., Aljaafari, H.A.: Promising evolution of biofuel generations subject review. Renew. Energy Focus **28**, 127–139 (2019)
6. Siwal, S.S., Zhang, Q., Devi, N., Saini, A.K., Saini, V., Pareek, B., et al.: Recovery processes of sustainable energy using different biomass and wastes. Renew. Sustain. Energy Rev. **150**, 111483 (2021)
7. Clifton-Brown, J., Hastings, A., von Cossel, M., Murphy-Bokern, D., McCalmont, J., Whitaker, J., et al.: Perennial biomass cropping and use: Shaping the policy ecosystem in European countries. GCB Bioenergy (2023)
8. Janiszewska, D., Żurek, G., Martyniak, D., Bałęczny, W.: Lignocellulosic biomass of C3 and C4 perennial grasses as a valuable feedstock for particleboard manufacture. Materials **15**(18), 6384 (2022)
9. Koul, B., Yakoob, M., Shah, M.P.: Agricultural waste management strategies for environmental sustainability. Environ. Res. **206**, 112285 (2022)
10. Saini, J.K., Saini, R., Tewari, L.: Lignocellulosic agriculture wastes as biomass feedstocks for second-generation bioethanol production: Concepts and recent developments. 3 Biotech. **5**, 337–53 (2015)

11. Meyer, A., Ehimen, E., Holm-Nielsen, J.: Future European biogas: Animal manure, straw and grass potentials for a sustainable European biogas production. Biomass Bioenergy **111**, 154–164 (2018)
12. Mohammed, N.I., Kabbashi, N., Alade, A.: Significance of agricultural residues in sustainable biofuel development. Agric. Waste Residues, 71–88 (2018)
13. Lee, S.Y., Sankaran, R., Chew, K.W., Tan, C.H., Krishnamoorthy, R., Chu, D.-T., et al.: Waste to bioenergy: A review on the recent conversion technologies. BMC Energy **1**(1), 1–22 (2019)
14. Alizadeh, P., Tabil, L.G., Mupondwa, E., Li, X., Cree, D.: Technoeconomic feasibility of bioenergy production from wood sawdust. Energies **16**(4), 1914 (2023)
15. Maier, J.M., Sowlati, T., Salazar, J.: Life cycle assessment of forest-based biomass for bioenergy: A case study in British Columbia, Canada. Resour. Conserv. Recycl. **146**, 598–609 (2019)
16. Malode, S.J., Prabhu, K.K., Mascarenhas, R.J., Shetti, N.P., Aminabhavi, T.M.: Recent advances and viability in biofuel production. Energy Convers. Manag.: X. **10**, 100070 (2021)
17. Abreu, M., Silva, L., Ribeiro, B., Ferreira, A., Alves, L., Paixão, S.M., et al.: Low indirect land use change (ILUC) energy crops to bioenergy and biofuels—a review. Energies **15**(12), 4348 (2022)
18. Saleem, M.: Possibility of utilizing agriculture biomass as a renewable and sustainable future energy source. Heliyon, e08905 (2022)
19. Xiao, M.-Z., Sun, Q., Hong, S., Chen, W.-J., Pang, B., Du, Z.-Y., et al.: Sweet sorghum for phytoremediation and bioethanol production. J. Leather Sci. Eng. **3**, 1–23 (2021)
20. Almomani, F., Hosseinzadeh-Bandbafha, H., Aghbashlo, M., Omar, A., Joo, S.-W., Vasseghian, Y., et al.: Comprehensive insights into conversion of microalgae to feed, food, and biofuels: Current status and key challenges towards implementation of sustainable biorefineries. Chem. Eng. J. **455**, 140588 (2023)
21. Ali, S.S., Mastropetros, S.G., Schagerl, M., Sakarika, M., Elsamahy, T., El-Sheekh, M., et al.: Recent advances in wastewater microalgae-based biofuels production: A state-of-the-art review. Energy Rep. **8**, 13253–13280 (2022)
22. Sarwer, A., Hamed, S.M., Osman, A.I., Jamil, F., Al-Muhtaseb, AaH., Alhajeri, N.S., et al.: Algal biomass valorization for biofuel production and carbon sequestration: A review. Environ. Chem. Lett. **20**(5), 2797–851 (2022)
23. Prabakaran, P., Karthikeyan, S.: Algae biofuel: A futuristic, sustainable, renewable and green fuel for IC engines. Mater. Today: Proc. (2023)
24. Braud, L., McDonnell, K., Murphy, F.: Environmental life cycle assessment of algae systems: Critical review of modelling approaches. Renew. Sustain. Energy Rev., 113218 (2023)
25. Murugesan, P., Raja, V., Dutta, S., Moses, J., Anandharamakrishnan, C.: Food waste valorisation via gasification–A review on emerging concepts, prospects and challenges. Sci. Total. Environ., 157955 (2022)
26. Nanda, S., Berruti, F.: A technical review of bioenergy and resource recovery from municipal solid waste. J. Hazard. Mater. **403**, 123970 (2021)
27. Samoraj, M., Mironiuk, M., Izydorczyk, G., Witek-Krowiak, A., Szopa, D., Moustakas, K., et al.: The challenges and perspectives for anaerobic digestion of animal waste and fertilizer application of the digestate. Chemosphere **295**, 133799 (2022)
28. Jayakumar, M., Gebeyehu, K.B., Abo, L.D., Tadesse, A.W., Vivekanandan, B., Sundramurthy, V.P., et al.: A comprehensive outlook on topical processing methods for biofuel production and its thermal applications: Current advances, sustainability and challenges. Fuel **349**, 128690 (2023)
29. Greinert, A., Mrówczyńska, M., Szefner, W.: The use of waste biomass from the wood industry and municipal sources for energy production. Sustainability. **11**(11), 3083 (2019)
30. Ilari, A., Duca, D., Boakye-Yiadom, K.A., Gasperini, T., Toscano, G.: Carbon footprint and feedstock quality of a real biomass power plant fed with forestry and agricultural residues. Resources **11**(2), 7 (2022)
31. Hu, Y., Gong, M., Xing, X., Wang, H., Zeng, Y., Xu, C.C.: Supercritical water gasification of biomass model compounds: A review. Renew. Sustain. Energy Rev. **118**, 109529 (2020)

32. Hoang, A.T., Ong, H.C., Fattah, I.R., Chong, C.T., Cheng, C.K., Sakthivel, R., et al.: Progress on the lignocellulosic biomass pyrolysis for biofuel production toward environmental sustainability. Fuel Process. Technol. **223**, 106997 (2021)
33. Leong, W.-H., Lim, J.-W., Lam, M.-K., Uemura, Y., Ho, Y.-C.: Third generation biofuels: A nutritional perspective in enhancing microbial lipid production. Renew. Sustain. Energy Rev. **91**, 950–961 (2018)
34. Cui, X., Yang, J., Wang, Z., Shi, X.: Better use of bioenergy: A critical review of co-pelletizing for biofuel manufacturing. Carbon Capture Sci. & Technol. **1**, 100005 (2021)
35. Kowalski, Z., Kulczycka, J., Verhé, R., Desender, L., De Clercq, G., Makara, A., et al.: Second-generation biofuel production from the organic fraction of municipal solid waste. Front. Energy Res. **10**, 919415 (2022)
36. Abbasi, M., Pishvaee, M.S., Mohseni, S.: Third-generation biofuel supply chain: A comprehensive review and future research directions. J. Clean. Prod. **323**, 129100 (2021)
37. Maliha, A., Abu-Hijleh, B.: A review on the current status and post-pandemic prospects of third-generation biofuels. Energy Systems, 1–32 (2022)
38. Mat Aron, N.S., Khoo, K.S., Chew, K.W., Show, P.L., Chen, W.H., Nguyen, T.H.P.: Sustainability of the four generations of biofuels—a review. Int. J. Energy Res. **44**(12), 9266–9282 (2020)
39. Cavelius, P., Engelhart-Straub, S., Mehlmer, N., Lercher, J., Awad, D., Brück, T.: The potential of biofuels from first to fourth generation. PLoS Biol. **21**(3), e3002063 (2023)
40. Pandit, C., Banerjee, S., Pandit, S., Lahiri, D., Kumar, V., Chaubey, K.K., et al.: Recent advances and challenges in the utilization of nanomaterials in transesterification for biodiesel production. Heliyon (2023)
41. Al-Bawwat, AaK., Cano, A., Gomaa, M.R., Jurado, F.: Availability of biomass and potential of nanotechnologies for bioenergy production in Jordan. Processes **11**(4), 992 (2023)
42. Singh, N., Dhanya, B., Verma, M.L.: Nano-immobilized biocatalysts and their potential biotechnological applications in bioenergy production. Mater. Sci. Energy Technol. **3**, 808–824 (2020)
43. Deng, W., Feng, Y., Fu, J., Guo, H., Guo, Y., Han, B., et al.: Catalytic conversion of lignocellulosic biomass into chemicals and fuels. Green Energy & Environ. (2022)
44. Liang, C., Wang, Q., Wang, W., Lin, C.S.K., Hu, Y., Qi, W.: Enhancement of an efficient enzyme cocktail from Penicillium consortium on biodegradation of pretreated poplar. Chem. Eng. J. **452**, 139352 (2023)

Study of the Evolution of Ozone Concentration as a Function of Nitrogen Monoxide and Dioxide in Ambient Air

Benchrifa Mohammed, Jamal Mabrouki, Khadija E. L. Moustaqim, and Driss Hmouni

Abstract The evolution of ozone as a function of nitrogen dioxide is a complex subject that requires a thorough understanding of the mechanisms of ozone formation. When nitrogen dioxide is present in the atmosphere, it reacts with volatile organic compounds (VOCs) under the influence of sunlight to form ground-level ozone. This chemical reaction is called photochemistry and can have adverse effects on human health and the environment. However, it is important to note that the relationship between ozone and nitrogen dioxide is not linear. Indeed, at low concentrations of nitrogen dioxide, ozone can be degraded, while at higher concentrations, ozone can be formed more quickly. Thus, to understand the evolution of ozone as a function of nitrogen dioxide, it is necessary to take into consideration other factors such as temperature, humidity and the presence of other atmospheric pollutants. Ultimately, we must all work together to promote sustainable lifestyles and reduce our impact on the environment in order to preserve our planet for future generations.

Keywords Evolution of ozone · Nitrogen dioxide · Air quality · Photochemistry

B. Mohammed (✉)
Laboratory of Solar Energy and Environment, Mohammed V University in Rabat, Faculty of Sciences, B.P. 1014, Rabat, Morocco
e-mail: simobenchrifa@gmail.com

J. Mabrouki
Laboratory of Spectroscopy, Molecular Modeling, Materials, Nanomaterial, Water and Environment, CERNE2D, Mohammed V University in Rabat, Faculty of Science, Avenue Ibn Battouta, BP1014 Rabat, Agdal, Morocco

B. Mohammed · K. E. L. Moustaqim · D. Hmouni
Improvement and Valuation of Plant Resources, Faculty of Sciences, Ibn Tofaïl University—KENITRA-University Campus, 14000 Kenitra, Morocco

J. Mabrouki and M. Azrour (eds.), *Integrated Solutions for Smart and Sustainable Environmental Conservation*, Studies in Systems, Decision and Control 527,
https://doi.org/10.1007/978-3-031-55787-3_5

1 Introduction

Atmospheric pollution is a growing concern for citizens, journalists, politicians, researchers and governments. It is a topical issue that has adverse effects on the environment, and this environment of land, water, plants and air is a very important whole for our well-being and for the balance of nature. In recent years, the assessment of air pollution, and more specifically dust pollution, has been considered a veritable revolution with a direct impact on human beings. The presence of smoke, ash and sulphurous gases in the atmosphere has long been recognized as costly and dangerous. Today, it's accepted that air pollution is more than just a minor inconvenience and can endanger human health. Severe pollution episodes have been shown to increase lethality, and separate research has suggested that there may be a link between more or less regular exposure to a polluted atmosphere and the onset of diseases such as acute bronchitis and primary lung cancer [1].

Atmospheric pollution can be of biogenic or anthropogenic origin. The main natural sources are wind erosion on the Earth's surface, ocean erosion, fires, biogenic gas emissions from ecosystems and volcanic eruptions. Anthropogenic emissions are of more diverse origins. They are mainly generated by the combustion of natural fossil resources (coal, gas, oil) or biomass, and by the evaporation of solvents and other volatile products. In the case of air pollution, the European Environment Agency defines pollution more precisely, taking into account the impact on human health: "Air pollution includes the presence of certain pollutants in the atmosphere at levels sufficient to affect human health, the environment and cultural heritage". Atmospheric pollutants include nitrogen monoxide and dioxide. The study of ozone concentrations as a function of NO and also NO_2 in ambient air is an important subject for understanding air quality and its impact on health and the environment. These three atmospheric compounds are closely linked and interact in the atmosphere.

Ozone (O_3) is a second type of pollutant, produced when NO_x and NO_2 react with volatile organic compounds (VOCs) in the form of sunlight. These chemical reactions take place in the atmosphere, generally near the Earth's surface. Power stations and industrial plants are the main sources of nitric oxide (NO) and NO_2. These gases also react with other atmospheric compounds, contributing to the formation of ground-level ozone [1]. When nitric oxide (NO) is emitted into the atmosphere, it reacts rapidly with O_3, destroying it and forming NO_2. NO_2 can then be converted to ozone by a photochemical reaction in sunlight. Consequently, high concentrations of nitric oxide promote ozone formation. In this chapter, we will study the evolution of ozone concentration as a function of nitrogen monoxide and dioxide in ambient air.

2 Material and Methods

2.1 *Ozone (O_3)*

Ozone is a highly reactive, unstable compound made up of three oxygen atoms. It is naturally present in the atmosphere. Ozone is formed by the photodissociation of oxygen molecules (O_2) in the upper atmosphere, more precisely in the stratosphere, which releases an oxygen atom (O). This then combines with another oxygen molecule (O_2) to form an ozone molecule (O_3).

The resulting ozone forms the ozone layer that protects life on earth by filtering out the sun's ultraviolet rays [2].

In the troposphere, ozone is formed differently. This time, the O atoms come from the photolysis of nitrogen dioxide (NO_2) molecules. However, low-altitude O_3 is considered a pollutant because of its oxidizing capacity to adversely affect health, vegetation and materials. Figure 1 shows the distribution of ozone, demonstrating the presence of "good" ozone at 15–50 km in the stratosphere, while "bad" ozone is found near the surface.

2.2 *Nitrogen Dioxide (NO_2)*

This is a suffocating reddish-brown toxic gas with a pungent odor [3]. It is the precursor to the industrial production of nitrate HNO_3, and is a major pollutant of the earth's atmosphere, produced by internal combustion engines and power plants; as such, it is responsible for the eutrophication of unsulfated acid rain (NO_2 combines

Fig. 1 Ozone vertical distribution

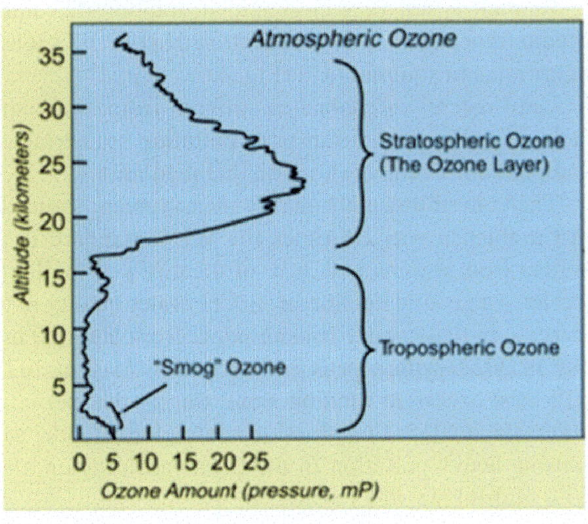

with ozone in anthropogenic tropospheric rain to form highly water-soluble nitrates) [3]. Nitric acid, which is formed by hydration of nitrogen dioxide, is also produced by nitrogen dioxide.

$$3NO_2 + H_2O \rightarrow 2\,HNO_3 + NO$$

Nitrogen dioxide (NO_2) has an unpleasant, pungent odor. Nitric acid and organic nitrates are formed when it reacts with air as an oxidizing agent. As well as contributing to photochemical pollution, it is involved in atmospheric reactions that produce ozone. Nitrogen dioxide is also a precursor of nitrates, which increase the quantity of respirable particles in the air. UV irradiation easily dissociates it from water.

2.3 Geographical Location of the Study Area

Casablanca is located in the northwest of the Casablanca-Settat region, and the Province of Casablanca currently has a population of 3,343,642. The province covers an area of 180 km^2 and is divided into two urban communes. Casablanca's geographic coordinates are 33° 33′ north latitude, 7° 35′ west longitude, bordering the Atlantic Ocean to the west, the province of Mohammedia to the north, the Atlantic Ocean to the west and bordering the province of Nouaceur. Benslimane County to the east and Benslimane County to the south [8] (Fig. 2).

Casablanca enjoys a very mild Mediterranean climate. Affected by the cold Atlantic current, the temperature in this region is the mildest in Morocco, with little seasonal variation, averaging 13 °C from January to 24 °C in August [9]. Here are the average temperatures (Fig. 3).

Rainfall is not very commercial, totaling 425 mm per year, and following the Mediterranean pattern, is concentrated between October and April, whereas it almost never rains in summer [10] (Fig. 4).

Until recently, Casablanca suffered from environmental consequences due to a lack of quality public transport, polluting industrial activities, limited green spaces and poor enforcement of municipal regulations [11].

There are thirteen air quality measurement stations in Casablanca. Air pollution in Casablanca was 2.5 times the WHO standard in 2014, 16% of the population suffers from asthma and 30% of the city's air pollution is caused by traffic. There has been a general decline in surface water quality in the Casablanca region. A 30% increase in daily water consumption, from 500,000 m^3/day in 2011 to 650,000 m^3/day in 2022, with a peak of 740,000 m^3/day in 2022. A total of 600,000 people will have access to running water during this period. The treatment rate has been 100% since 2015. Construction contractors plunder sand from Casablanca beaches, causing heavy pollution In the Casablanca region, the Oum Errabiâ wadi supplies 50% of drinking water needs.

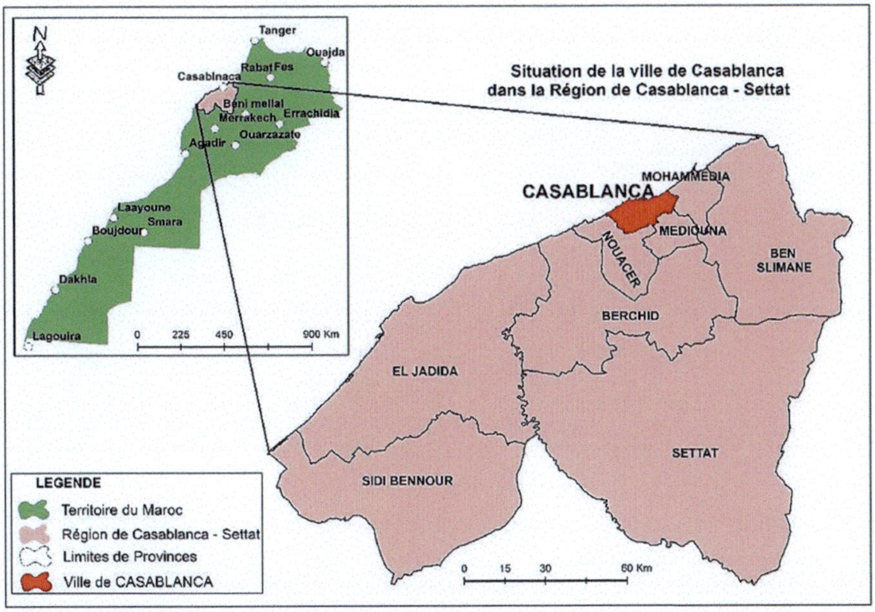

Fig. 2 Geographical location of Casablanca

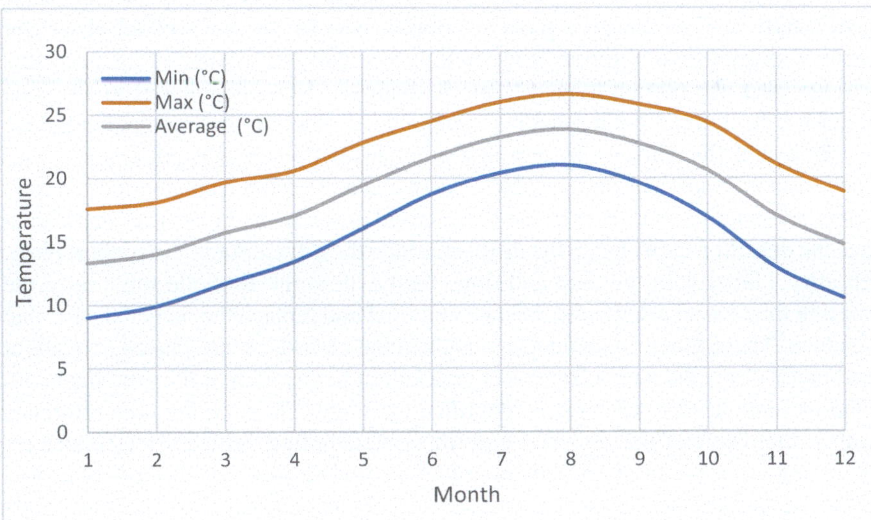

Fig. 3 Average temperatures (1991–2020) in Casablanca

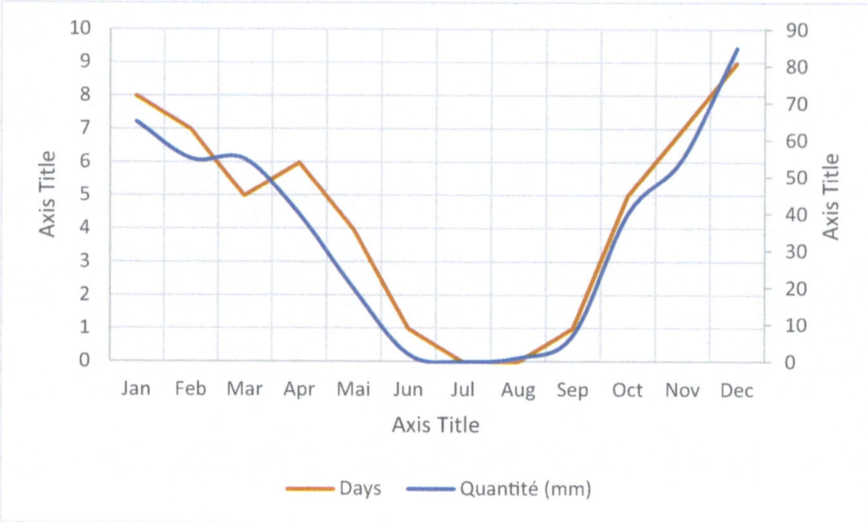

Fig. 4 Average rainfall in Casablanca

2.4 *Measurement Techniques Used*

In this study, a Radiello-type passive sampler was In the form of an absorbent cartridge inserted into a porous body held horizontally by triangular supports,.the cartridge used was 60 mm long, 4.8 mm in diameter, NO_2 code name166, O_3 code name 172. To trap the photosensitive compound, a blue microporous polyethylene diffuser with a diameter of 16 mm and a height of 60 mm is used; it has a wall thickness of 1.7 mm, a porosity of 25 ± 5 μm and a diffusion path length of 18 mm. A polycarbonate support plate (code 121) is used to screw on the diffuser body. First place the detection cartridge in the diffuser, screw the diffuser onto the support plate, then affix a label with the date and time, then expose the Radiello tube, once the sampling time has elapsed, note the date and time and replace the recycled cartridge in its tube. These tubes, once fitted with adsorbent cartridges, are placed at a height of around 3–4 m. We have taken every precaution to ensure that the pipeline is not stolen or vandalized. Each tube is identified by a label showing the tube reference number, the date and the start and end times of sampling, which are essential for calculating exposure duration. This sampling method has several advantages: a large number of measurements can be carried out simply and quickly, and the compounds present in the sample can be identified without a predefined list.

The analytical principle of nitrogen dioxide NO_2 is to convert it completely to the nitrite ion, then complex it in the presence of sulfonamide (SA) and N-(1 naphthyl) ethylenediamine dihydrochloride (NEDA), followed by spectrophotometric quantification. After collecting the conditioned tubes from the fridge, pour 5 ml of distilled water into each tube and shake for three minutes. Next, remove 0.5 ml of the extract from the column, place in a stoppered glass tube and add 5 ml of SA reagent. The tubes were then recapped and shaken. After 5 min, 1 ml NEDA was added and the mixture shaken. A pink-purple color appeared, and the degree of color development was proportional to the amount of NO_2 captured. Ten minutes after the addition of NEDA, the absorbance of the samples was measured at 537 nm by UV–Visible spectrophotometry. An ozone analysis is performed by first preparing a solution of MBTH hydrazone chloride; weigh out 5 g of MBTH, pour it into a 1-L bottle, add 1 L of demineralized water, and finally add 5 ml of concentrated sulfuric acid, then remove the filter element from the plastic tube, remove the PTFE cap, and pour the contents into the same plastic tube. Shake the tube well after adding 5 ml of MTBH solution. After one hour, shake occasionally. In a second tube, filter the mixture and measure the absorbance at 430 nm of the filtered solution. The solution remains closed for several days and remains the same color.

2.5 Data Processing Technique

STATISTICA statistical software was used to process the measured data. STATISTICA is a statistical and data analysis software package developed by TIBCO Software [12]. It provides extensive functionality for data exploration, visualization and statistical analysis. Here are some of Statistica's key features: Data exploration: Statistica lets you explore data using advanced visualization techniques such as interactive graphs, scatter plots, correlation matrices and more. It also provides tools for summarizing and organizing data. Statistical analysis: the software provides a variety of statistical techniques, such as linear and non-linear regression, analysis of variance (ANOVA), analysis of longitudinal data, survival analysis, clustering methods, etc. It also enables you to carry out tests on the results of your analyses. It can also be used to perform statistical tests to verify hypotheses and evaluate relationships between variables [13]. Predictive modeling: Statistica provides predictive modeling capabilities for building statistical models. It enables the development of regression models, decision trees, neural networks, classification methods and more. These models can be used to make predictions and data-driven decisions.

3 Results and Discussion

Spectrophotometric analysis enabled us to assess the level of NO_2 pollution in Casablanca. Table 1 describes the nitrogen dioxide results measured during 2011 and 2012.

For ozone, Table 2 describes the results measured during 2011 and 2012.

After completing the study, we graphically illustrate in Figs. 5 and 6 the difference in evolution of ozone and nitrogen dioxide concentrations during 2011 and 2012 measure in Casablanca.

Table 1 Statistical description of NO_2

Variable	Descriptive statistics								
	Valid N	Median	Mode	Frequency mode	Minimum	Maximum	25000th percentile	75000th percentile	Geometric mean
2011	12	24,00000	27,00000	3	5,000000	30,00000	17,50000	27,00000	19,84390
2012	12	20,00000	14,00000	2	7,000000	36,00000	14,00000	28,50000	18,90261
	Harmonic mean	Std.Dev.	Variance	Average deviation	Range	Quartile rassnge	Skewness	Kurtosis	Sum
2011	16,90428	7,617265	58,02273	5,958333	25,00000	9,50000	−1,12062	0,62674	261,0000
2012	16,80560	9,179605	84,26515	7,750000	29,00000	14,50000	0,20516	−1,12200	251,0000

Table 2 Statistical description of O_3

Variable	Descriptive statistics (Spreadsheet1_(Recovered))							
	Median	Mode	Frequency mode	Minimum	Maximum	25000th percentile	75000th percentile	Geometric mean
2011	298,5000	Mode		263,0000	315,0000	278,5000	307,5000	293,0144
2012	297,5000	multiple		273,0000	331,0000	278,0000	317,50000	299,0391

	Harmonic Mean	Std.Dev.	Variance	Average deviation	Range	Quartile range	Skewness
2011	292,5213	17,49026	305,9091	14,91667	52,00000	29,00000	−0,433250
2012	298,3339	21,62543	467,6591	18,91667	58,00000	39,50000	0,206432

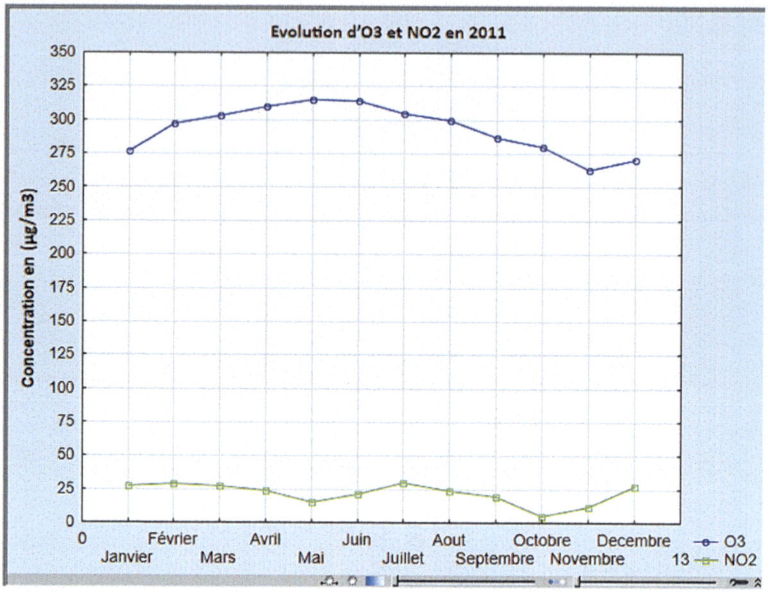

Fig. 5 O_3 and NO_2 trends in 2011

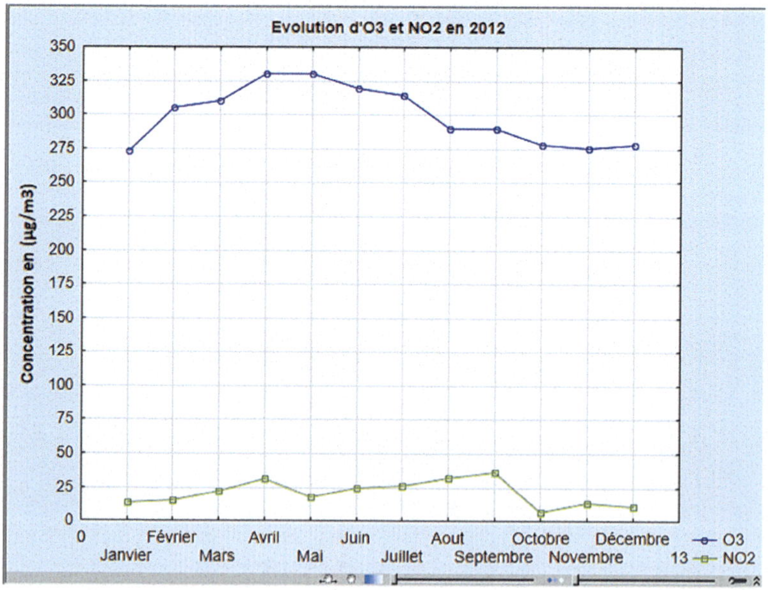

Fig. 6 O_3 and NO_2 trends in 2012

According to Tables 1 and 2, the statistical study described the concentrations measured during 2011 and 2012, for nitrogen dioxide we note that the standard deviation or standard deviation quantified the average dispersion of values in relation to the mean, we note a convergence in both years (7.617265 in 2011 and 9.179605 in 2012) and for ozone (17.49026) in 2011 and 21.62543 in 2012). About the median, it's a measure of central position that divides the data into two equal parts, it represents the value in the middle of the data set when sorted in ascending or descending order, for nitrogen dioxide we note (24.00 in 2011 and 20.00 in 2012) and for ozone (298.50 in 2011 and 297.50 in 2012).

Figure 5 shows that the O_3 curve rises from January to May, unlike NO_2, which initially rose from January to February but then began to fall until May, after which it began to rise to around 25 ug/m^3, while O_3 began to fall, reaching its minimum value in November at around 262, 5 ug/m^3 and is beginning to rise again, while NO^2 values underwent a series of ups and downs as it began to rise from May to July and then declined again until it reached more or less zero and is rising again. In 2012, both variables rose until April, when the O_3 value remained stable until the following month, then fell towards October and remained more or less constant, while the NO_2 value fell from April to May, then rose slowly until it reached almost zero in October, then rose again. It has been shown that the presence of nitrogen dioxide can have a significant impact on ozone evolution, when NO_2 is exposed to sunlight, it breaks down into nitrogen radicals (NO_x) which are reactive species and play a crucial role in ozone formation, NO_2 levels vary from 5 μg/m^3 to 36 μg/m^3, resulting in a notable increase in ozone evolution ranging from 263 μg/m^3 to 331 μg/m^3. When NO and NO_2 concentrations rise, ozone production is stimulated, as more precursors are available. However, above a certain threshold, an excessive increase in NOx leads to an increase in ozone destruction through competitive chemical reactions. In Casablanca, nitrogen dioxide pollution levels are high, as evidenced by the deterioration in traffic on major roads and the very high population density. There are two main sources of nitrogen dioxide: Nitrogen dioxide is a recognized indicator of urban pollution from external sources. The majority comes from forestry sources (boiler rooms, power stations) and mobile sources (vehicle engines). Gas heaters, gas stoves and other gas-burning appliances produce nitrogen dioxide indoors. In addition to tobacco smoke, it comes from the environment.

In fact, ozone is a pollutant whose highest levels are recorded in the early afternoon, when the main conditions for ozone peaks are: Sunlight at its highest level, which directly causes the formation of ozone precursors (nitrogen oxide and volatile organic compounds) in urban areas. A natural ozone concentration varies between 20 and 45 parts per billion, depending on season and altitude, and is at its lowest at night, when there is no photochemical action. In fact, there are two categories of ozone levels: natural and artificial. Natural ozone occurs naturally in the troposphere; some natural ozone comes from the stratosphere, and some is produced by the reaction of methane with large quantities of moisture. Due to the small size of the lower layer, the NO produced by nighttime air does not mix with the inversion layer.

Several geographical, seasonal and meteorological factors influence the interaction between nitrogen dioxide and ozone. In addition to nitric oxide (NO), nitrogen dioxide (NO_2) is another ozone precursor. Cars, power stations and industrial plants emit these precursors when they burn fossil fuels. Ozone is formed when nitric oxide and nitrogen dioxide react with volatile organic compounds in sunlight. In summer, ozone concentrations tend to be higher due to this chemical reaction, which occurs mainly during periods of high light levels and heat. However, there is a complex relationship between nitric oxide, nitrogen dioxide and ozone concentrations and several other factors. Precursor availability, meteorological conditions and atmospheric dispersion patterns all play a role. High concentrations of nitric oxide and nitrogen dioxide may inhibit ozone formation in some cases, while promoting it in others. NO_2 concentrations are generally higher in winter, due to emissions from heating systems and increased vehicle traffic in cold weather. However, ozone levels are generally lower in winter due to the lower intensity of sunlight required for ozone formation. Consequently, the relationship between ozone and NO_2 is reversible throughout the year. Higher temperatures generally accelerate the chemical reactions involved in ozone formation. As a result, during the hottest months of the year, ozone levels can rise even when nitrogen dioxide concentrations are not high. Ozone production is closely linked to meteorological conditions, particularly temperature and sunshine. Prolonged exposure to high temperatures and sunlight accelerates chemical reactions, resulting in the formation of ozone from precursors such as nitrogen dioxide. Winds can carry nitrogen dioxide over long distances, affecting ozone concentrations in different regions. Regional emissions of nitrogen dioxide can also be transported to other regions, increasing ozone concentrations far from the source of emission.

4 Conclusion

An air quality study conducted in Casablanca used a passive sampling method, which was effective in measuring ozone (O_3) and nitrogen dioxide (NO_2) concentrations in ambient air. To collect representative samples of air pollution, passive sensors are exposed to ambient air over a period of time. According to the results of this study, the maximum NO_2 concentration measured was 36 g/m^3, while the maximum O_3 concentration reached 331 g/m^3. Casablanca has high levels of air pollution, as these figures show. During the period in which the measurements were taken, the results of this study are valid only for that period. There may be variations in meteorological conditions, so the factors favoring ozone formation may not be the same as those favoring nitrogen dioxide. When nitrogen oxide (NO_x) levels are high in Casablanca, ozone undergoes a chemical reaction called titration when nitric oxide (NO) is present. As a result of this reaction, nitrogen dioxide (NO_2) is produced and ozone levels are reduced. When NO_x levels are moderate and volatile organic compound (VOC) levels are sufficiently high, ozone formation can occur. As a result, ozone levels can rise, underlining the importance of considering how pollution interacts with each other.

References

1. Benchrifa, M., Mabrouki, J., Tadili, R.: Development of a new solar system for heating and cooling an agricultural greenhouse. Eng. Proc. **5** (2023). https://doi.org/10.3390/xxxxx Published: 17 May 2023
2. Benchrifa, M., et al.: Modelling and simulating the effect of irradiation variation on the behavior of a photovoltaic cell and its influence on the maximum power point. In: Mabrouki, J., Mourade, A., Irshad, A., Chaudhry, S. (eds.) Advanced Technology for Smart Environment and Energy. Environmental Science and Engineering. Springer, Cham. https://doi.org/10.1007/978-3-031-25662-2_9
3. Benchrifa, M., Elouardi, M., Fattah, G., et al.: Identification, Simulation and modeling of the main power losses of a photovoltaic installation and use of the internet of things to minimize system losses. In: Advanced Technology for Smart Environment and Energy. Springer International Publishing, Cham, pp. 49–60 (2023). https://doi.org/10.1007/978-3-031-25662-2_4
4. Benchrifa, M., Mabrouki, J., et Tadili, R.: Estimation of global irradiation on horizontal plane using artificial neural network. In: Artificial Intelligence and Smart Environment: ICAISE'2022. Springer International Publishing, Cham, pp. 395–400 (2023). https://doi.org/10.1007/978-3-031-26254-8_56
5. Benchrifa, M., Mabrouki, J., Elouardi, M., Azrour, M., Tadili, R.: Detailed study of dimensioning and simulating a grid-connected PV power station and analysis of its environmental and economic effect, case study. Model. Earth Syst. Environ., 1–9 (2022). https://doi.org/10.1007/s40808-022-01457-9
6. Benchrifa, M., Mabrouki, J.: Simulation, sizing, economic evaluation and environmental impact assessment of a photovoltaic power plant for the electrification of an establishment.Adv. Build. Energy Res., 1–18 (2022). https://doi.org/10.1080/17512549.2022.2096693
7. Benchrifa, M., Essalhi, H., Tadili, R., Nfaoui, H.: Estimation of daily direct solar radiation for rabat. In: Sayigh, A. (ed.) Sustainable Energy Development and Innovation. Innovative Renewable Energy. Springer, Cham (2022). https://doi.org/10.1007/978-3-030-76221-6_69
8. Benchrifa, M., Tadili, R., Idrissi, A., Essalhi, H., Mechaqrane, A.: Development of new models for the estimation of hourly components of solar radiation: tests, comparisons, and application for the generation of a solar database in Morocco. Int. J. Photoenergy **2021**(8897818), 16 (2021). https://doi.org/10.1155/2021/8897818
9. Benchrifa, M., Essalhi, H., Tadili, R., Bargach, M.N., Mechaqrane, A.: "Development of a daily databank of solar radiation components for Moroccan Territory". Int. J. Photoenergy **2019**(Article ID 6067539), 20 (2019). https://doi.org/10.1155/2019/6067539.
10. Mabrouki, J., Benchrifa, M., Ennouhi, M., et al.: Geographic information system for the study of water resources in Chaâba El Hamra, Mohammedia (Morocco). In: Artificial Intelligence and Smart Environment: ICAISE'2022. Springer International Publishing, Cham, pp. 469–474 (2023). https://doi.org/10.1007/978-3-031-26254-8_67
11. Essalhi, H., Benchrifa, M., Tadili, R.: Effect of natural and forced ventilation on drying of Kiwi in an Indirect Solar Dryer. In: Sayigh, A. (ed.) Sustainable Energy Development and Innovation. Innovative Renewable Energy. Springer, Cham (2022). https://doi.org/10.1007/978-3-030-76221-6_24
12. Idrissi, A., Benchrifa, M., Tadili, R., Naciri, M., El Madrab, S.: Experimental investigation of solar still with corrugated absorber surface for domestic desalination use. In: Sayigh, A. (ed.) Sustainable Energy Development and Innovation. Innovative Renewable Energy. Springer, Cham (2022). https://doi.org/10.1007/978-3-030-76221-6_26
13. Essalhi, H., Benchrifa, M., Tadili, R., Bargach, M.N.: Experimental and theoretical analysis of drying grapes under an indirect solar dryer and in open sun. Innov. Food Sci. & Emerg. Technol. **49**, 58–64 (2018). ISSN 1466-8564, https://doi.org/10.1016/j.ifset.2018.08.002

14. Elouardi, M., Fattah, G., Benchrifa, M., et al.: The evaluation of the valorization of cannabis residues for the production of energy by combustion. In: Advanced Technology for Smart Environment and Energy. Springer International Publishing, Cham, pp. 247–256 (2023). https://doi.org/10.1007/978-3-031-25662-2_20
15. Fattah, G., Elouardi, M., Benchrifa, M., et al.: Modeling of the coagulation system for treatment of real water rejects. In: Advanced Technology for Smart Environment and Energy. Springer International Publishing, Cham, pp. 161–171 (2023). https://doi.org/10.1007/978-3-031-25662-2_14
16. Azoulay, K., Bencheikh, I., Benchrifa, M., et al.: Adsorption of inorganic and organic pollutants in urban wastewater treatment using pine wood activated carbon. In: Advanced Technology for Smart Environment and Energy. Springer International Publishing, Cham, pp. 221–231 (2023). https://doi.org/10.1007/978-3-031-25662-2_18
17. Ihoume, I., Tadili, R., Arbaoui, N., Benchrifa, M., Idrissi, A., & Daoudi, M.: Developing a TinyML-oriented deep learning model for an intelligent greenhouse microclimate control from multivariate sensed data. In: Intelligent Sustainable Systems: Selected Papers of WorldS4 2022, vol. 2, pp. 283–291. Springer Nature Singapore, Singapore (2023). https://doi.org/10.1007/978-981-19-7663-6_27
18. Ihoume, I., Tadili, R., Arbaoui, N., Benchrifa, M., Idrissi, A., Daoudi, M.: Developing a multi-label tinyML machine learning model for an active and optimized greenhouse microclimate control from multivariate sensed data. Artif. Intell. Agric. (2022). https://doi.org/10.1016/j.aiia.2022.08.003
19. Ihoume, I., Tadili, R., Arbaoui, N., Bazgaou, A., Idrissi, A., Benchrifa, M., Fatnassi, H.: Performance study of a sustainable solar heating system based on a copper coil water to air heat exchanger for greenhouse heating. Sol. Energy **232**, 128–138 (2022). https://doi.org/10.1016/j.solener.2021.12.064
20. Daoudi, M., Mou, A.A.S., Idrissi, A., Ihoume, I., Arbaoui, N., Benchrifa, M.: New approach to prioritize wind farm sites by data envelopment analysis method: A case study. Ocean Eng. **271**, 113820 (2023). https://doi.org/10.1016/j.oceaneng.2023.113820
21. Anouzla, A., et al.: Optimization of performance the waste in treatment of textile wastewater using response surface methodology. In: Mabrouki, J., Mourade, A., Irshad, A., Chaudhry, S. (eds.) Advanced Technology for Smart Environment and Energy. Environmental Science and Engineering. Springer, Cham (2023). https://doi.org/10.1007/978-3-031-25662-2_12
22. Elouardi, M., Zair, T., Mabrouki, J., Fattah, G., Benchrifa, M., Qisse, N., Belghiti, M.A.E.: A review of botanical, biogeographical phytochemical and toxicological aspects of the toxic plants in Morocco. Toxicol. Anal. Clin. (2022). https://doi.org/10.1016/j.toxac.2022.04.003

Wastewater as a Valuable Resource: Implications for Human Health and Ecosystems

Karima Azoulay, Imane Bencheikh, and Jamal Mabrouki

Abstract The increasing demand for water worldwide has led to a significant rise in wastewater production, resulting in a higher pollutant load. Unfortunately, in most developing countries, wastewater is discharged directly into the environment without adequate treatment, causing adverse effects on human health, economic productivity, freshwater resources, and ecosystems. Despite its importance in the water management cycle, wastewater is often viewed as a burden or ignored altogether. This approach poses serious implications for community well-being and individual livelihoods. Neglecting wastewater management could also hinder progress in achieving the 2030 Agenda for Sustainable Development. However, there is a growing recognition of the potential of wastewater as an alternative water source. This has shifted the paradigm from simply treating and disposing of wastewater to focusing on reuse, recycling, and resource recovery. Wastewater is no longer viewed as a problem requiring a solution; instead, it is seen as an integral part of addressing contemporary challenges. It presents an opportunity to serve as a cost-effective and sustainable source of energy, nutrients, and valuable by-products. By extracting these resources from wastewater, we can extend the positive impacts beyond just human health and environmental preservation. Embracing a circular economy approach, which balances economic development with natural resource protection and sustainability, recognizes wastewater as a widely available and valuable resource. By taking action now, there is an optimistic outlook for transforming the perception of wastewater and maximizing its potential for positive contributions to society and the environment.

Keywords Water resources · Wastewater · Pollution · Groundwater · Sustainable management · Ecosystems

K. Azoulay (✉) · I. Bencheikh · J. Mabrouki
Mohammed V University, Rabat, Morocco
e-mail: karima.azoulay@um5r.ac.ma

J. Mabrouki and M. Azrour (eds.), *Integrated Solutions for Smart and Sustainable Environmental Conservation*, Studies in Systems, Decision and Control 527,
https://doi.org/10.1007/978-3-031-55787-3_6

1 Introduction

The preservation and recovery of the quality of water resources are the major challenges for our society, both in the short term and for future generations. This is the objective set by the European Community which, through the Water Framework Directive of 2000 (WFD), obliges its member states to restore and protect the quality of their ground and surface water resources, with the first deadline of 2015, and continues at the national level through the MPWRDM (Master Plan for Water Resources Development and Management) [1]. The United Nations Interagency Group on Water in Africa (UN Water-Africa) is developing a mechanism for monitoring and evaluating the implementation of the African Water Vision 2025 [2]. In France, the commitments of the Grenelle Environment Forum have set the objective of achieving good ecological status for more than two-thirds of water bodies and for at least 90% by 2021, i.e. water of sufficient quality and quantity to ensure the sustainable functioning of ecosystems and satisfy all uses. Environmental pollution is classified into two types, anthropogenic on one hand, which corresponds to the discharge of substances into the environment as a result of human activity, and natural on the other hand, which is due to substances that are naturally found in water [3]. There is a wide variety of anthropogenic pollution sources, characterized by different substances and discharge modes, the main sources being agriculture, which is diffused over the territory, industries, which are the source of very diversified and often localized discharges, and domestic human activity via wastewater discharges [4]. Although groundwater is often less directly exposed to pollution than surface water, it can persist over time and it is very difficult to remedy contamination in the groundwater environment. Groundwater also provides the base flow for most rivers and can influence the quality of surface water. Groundwater is also widely used for drinking water supply, industry and agriculture, and its quality is therefore essential for human health and the proper functioning of economic activities [5]. Taking into account the surface water-groundwater continuum is therefore essential in the fight against pollution and the protection of water resources. Thus, the protection of water resources and environments requires a better knowledge of the sources and vectors of pollution in order to be able to take adequate measures to fight against this pollution within the framework of sustainable management of resources [6]. One of the major environmental challenges today is to be able to discriminate between the different sources of pollution, whether for a given element or molecule, i.e. to be able to identify components from domestic wastewater discharges, industrial discharges or agricultural activities. Among the identified vectors of pollution, domestic wastewater discharges from wastewater treatment plants or autonomous sanitation, constitute punctual entry points of contaminants in surface waters and/ or towards groundwater [7]. The objective of this article is to define the impact of domestic and industrial wastewater on groundwater.

2 Water Resources

2.1 *Wastewater: Global Trends*

The treatment of municipal and industrial wastewater varies significantly across different income countries. On average, high-income countries treat around 70% of the wastewater they produce. However, this percentage drops to 38% in upper-middle-income countries and further decreases to 28% in lower-middle-income countries. In low-income countries, a mere 8% of wastewater undergoes any kind of treatment. These figures highlight the stark disparities in wastewater treatment practices worldwide, leading to the widely quoted estimate that more than 80% of the world's wastewater is likely discharged without any treatment.

In high-income countries, the motivation for implementing advanced wastewater treatment techniques is primarily driven by the need to maintain environmental quality and to create an alternative water source to address water scarcity issues. Despite these motivations, the discharge of untreated wastewater remains a common practice, particularly in developing countries. This is mainly due to the lack of adequate infrastructure, technical and institutional capacity, and financial resources necessary to establish and maintain efficient wastewater treatment systems.

It is crucial to address these challenges and invest in improving wastewater treatment practices, especially in developing nations, to protect the environment, conserve water resources, and safeguard public health. Sustainable solutions and international cooperation are essential to promote better wastewater management globally [8].

2.2 *Source of the Problem*

2.2.1 Sustainable Approaches to Municipal and Urban Wastewater Management

Municipal wastewater is a complex mixture of contaminants from various sources, including domestic, industrial, commercial, and institutional activities. Emerging pollutants, like commonly used drugs, pose environmental and health risks even at low concentrations. With rapid urban growth, the generation of municipal wastewater has significantly increased, posing challenges for water management. One approach involves utilizing treated wastewater and by-products for various purposes to promote sustainability and resource efficiency.

Informal settlements in the developing world, such as slums, face critical wastewater management issues. Lack of proper sanitation facilities and clean water access lead to unsanitary practices like open defecation. Women in slums face additional challenges related to finding safe and suitable places to use toilets. Addressing these challenges requires multifaceted solutions, including improved sanitation infrastructure, access to clean water, and hygiene awareness programs. Prioritizing the needs of vulnerable populations is crucial for inclusive and sustainable urban development, benefiting residents' well-being and safeguarding the environment from untreated wastewater effects [10].

2.2.2 Industry

Minimizing pollution volumes and toxicity at the source is the first crucial step in dealing with industrial pollutants. This involves using environmentally friendly raw materials and biodegradable chemicals from the concept phase through design, operation, and maintenance. Educating and training personnel on pollution management are also vital to effectively address potential pollution issues.

The second step to reduce environmental impact is maximizing water recycling within the industrial facility, thus minimizing the need for wastewater discharge. Small and Medium-sized Enterprises (SMEs) and the informal sector often discharge wastewater into municipal systems or the environment, leading to serious pollution problems. To comply with discharge regulations and avoid fines, industries may be required to treat their wastewater before discharge to municipal systems or surface water [11].

However, there are instances where industries might find it more economically attractive to pay fines rather than investing in treatment to meet regulations. To promote a more sustainable approach, industrial symbiosis presents a notable opportunity for the use and recycling of industrial wastewater. This involves cooperation between different plants located in close proximity to each other. Eco-industrial parks are an excellent example of this concept, where industries are strategically placed to leverage different wastewater streams and facilitate water and by-product recycling.

For SMEs, participating in industrial symbiosis can be a significant way to save on wastewater treatment costs while contributing to a more sustainable and environmentally responsible industrial operation. By implementing these strategies, industries can mitigate their impact on water resources and the environment, leading to more sustainable industrial practices [12].

2.2.3 Agriculture

Over the last five decades, there has been a significant expansion in irrigated areas, with their size more than doubling. Additionally, the total number of livestock has tripled, and inland aquaculture has increased by over twenty times. However, this growth in agricultural practices has also led to water pollution issues.

Agricultural water pollution occurs when excessive amounts of fertilizers and agrochemicals are applied, surpassing the crop's capacity to absorb them or when heavy rain washes these chemicals into water bodies. Efficient irrigation systems can help reduce water and fertilizer losses, contributing to better water management.

Apart from nutrient pollution, livestock production and aquaculture activities also release nutrients and other pollutants into the water. The agricultural sector is a significant source of various pollutants, including organic matter, pathogens, metals, and emerging contaminants.

In recent years, the emergence of new agricultural pollutants has become a concern. Substances like antibiotics, vaccines, growth promoters, and hormones used in livestock and aquaculture operations can be released into water bodies, further exacerbating the pollution problem.

However, if handled correctly and applied safely, domestic wastewater can serve as a valuable resource, providing water and nutrients for agricultural purposes. Proper treatment and application of domestic wastewater can offer a sustainable solution to meet agricultural water demands and reduce reliance on freshwater sources [13]. Water reuse for agriculture offers not only the advantage of enhancing food security but also brings significant benefits to human and animal health, leading to improved nutrition. This approach of utilizing municipal wastewater for irrigation is widely adopted in several regions, such as the Middle East, North Africa, Australia, the Mediterranean, China, Mexico, and the United States.

The practice of wastewater reuse has proven to be highly successful, particularly in urban and peri-urban areas. In these regions, wastewater is readily available, often at no cost, and there is a nearby market for agricultural products, creating a conducive environment for its implementation.

By repurposing wastewater for irrigation, this sustainable model promotes efficient water management and reduces the demand on freshwater resources, which is especially crucial in water-scarce regions. Moreover, the nutrients present in the treated wastewater can enrich the soil and contribute to the growth of healthier crops, thus improving the overall nutritional quality of agricultural products.

However, it is essential to emphasize the significance of proper wastewater treatment and safe application practices to ensure that potential health and environmental risks are minimized. When managed correctly, water reuse for agriculture can be a valuable strategy to address food and water challenges while simultaneously contributing to better human and animal health outcomes [14].

3 Wastewater Characteristics

The composition of urban and municipal wastewater can vary significantly across different regions and is influenced by various factors, including domestic water consumption and the level of commercial and industrial activities. The specific composition of wastewater is essential to consider when designing wastewater treatment systems. Table 1 provides a summary of selected parameters commonly found in wastewater [15].

In developed regions with a higher degree of industrialization, the ratio of Biochemical Oxygen Demand (BOD) to Chemical Oxygen Demand (COD) is likely

Table 1 Composition of raw wastewater in certain countries

Parameters	United states	France	Morocco	Pakistan	Jordan
COD (mg/L)	110–400	100–400	45	193–762	152
BOD (mg/L)	250–1000	300–1000	200	83–103	386
TSS (mg/L)	100–350	150–500	160	76–6258	ND
Total phosphorus (mg/L)	4–15	1–25	4–5	ND	36
Total potassium/total nitrogen (mg/L)	20–85	30–100	29	ND	28

to be lower compared to developing regions. This is because there is a higher proportion of industrial wastewater in developed areas, making the water less suitable for biological treatment processes. As a result, alternative or advanced treatment methods may be required.

Certain regions may experience elevated levels of inorganic substances in their wastewater, such as sulfates and alkalinity. These substances can impact the water's suitability for various uses even after treatment. For instance, high sulfate levels can lead to the production of hydrogen sulfide, which can cause corrosion in sewer systems and other infrastructure.

Understanding the specific composition of wastewater in a given region is crucial for designing effective and appropriate treatment strategies. By addressing the unique challenges posed by different wastewater characteristics, authorities can develop more efficient and environmentally friendly approaches to manage urban and municipal wastewater effectively.

Likewise, high alkalinity or water hardness can lead to the formation of scale deposits and impair the water's suitability for reuse, for example as process water. Table 1 shows some of the most important pollutants present in different wastewater sources [16].

4 Impact of Wastewater on Groundwater

Wastewater consists mainly of water, constituting about 99% of its composition, with the remaining 1% comprising suspended colloidal and dissolved solids. When untreated or inadequately treated, the discharge of wastewater can lead to various consequences, which can be categorized into three main groups [17]:

(i) Adverse impact on human health;

(ii) Adverse impact on the environment;

iii) Adverse impact on economic activities.

4.1 Influence of Urbanization on Wastewater Generation: Trends and Consequences

Urban areas around the world face major challenges from accelerated urban growth, changing home and work practices, and urban sprawl. This mini review looks at the sources and consequences of municipal wastewater, while exploring opportunities for wastewater reuse and recycling. The emergence of informal human settlements poses a growing threat to service delivery, exacerbated by extreme events, climate change and migration from conflict-prone areas. As the pattern of urbanization changes and inequality becomes more pronounced, the poor in some developed regions face challenges similar to those in developing regions [18].

Looking ahead to 2030, the global demand for energy and water resources is projected to increase by 40% and 50% respectively (UN-Habitat, 2016). A significant portion of this growth will occur in cities, necessitating innovative approaches to wastewater management. Interestingly, effective wastewater management can also offer solutions to other pressing challenges, such as sustainable food production and industrial development [19].

To address these complex issues, cities must adopt comprehensive and sustainable wastewater management strategies. Embracing innovative technologies and practices will be crucial to efficiently handle wastewater, reduce pollution, and make the most of water resources. Additionally, promoting equitable access to essential services and improving living conditions in informal settlements are critical steps toward achieving inclusive and resilient urban development.

As the global population continues to urbanize, proactive measures to manage wastewater and water resources effectively are vital for the well-being of communities, the environment, and overall sustainable development.

5 Solution

The ultimate objective of wastewater management is to control and regulate the diverse wastewater flows effectively. This management cycle can be broken down into four distinct phases.

5.1 Source Pollution Prevention: Minimizing Contaminants at the Origin

Water pollution control should prioritize wastewater prevention and reduction over traditional end-of-pipe treatment for sustainable and cost-effective environmental management. This approach involves enforcing regulations and controls to restrict the use of certain pollutants and limit their entry into wastewater. By addressing pollution at its source, the need for expensive remedial actions is reduced, saving resources.

Preventive measures are generally more cost-effective than cleanup efforts. By proactively reducing pollution, resources can be allocated efficiently to prevent environmental degradation. Regular monitoring and reporting of pollutant discharges and water quality are essential for effective pollution control. Accurate data allows for better problem definition, achievable targets, and evaluation of pollution reduction efforts [20].

In summary, a preventive approach to water pollution control is vital for safeguarding water resources and the environment. Focusing on regulations, technical measures, and monitoring helps achieve sustainable and cost-effective environmental management, leading to healthier ecosystems and improved water quality.

5.2 Wastewater Collection and Treatment

Apologies for the detailed response. In summary, centralized disposal of waterborne wastes through sewer systems remains the primary method for wastewater management. However, other sanitation alternatives, such as on-site systems, are suitable for rural areas. Large-scale centralized wastewater treatment systems may not always be the most viable option for urban water management, leading to the growing trend of de-centralized wastewater treatment systems, which offer localized solutions with various benefits. These systems are proving to be more adaptable and efficient, contributing to resource conservation and water security [21].

Low-cost sewage treatment facilities have become a preferred method for various neighborhoods, regardless of their income levels. These facilities offer several advantages over conventional treatment plants, including significantly lower capital costs (around 20–50% of conventional plants) and reduced operation and maintenance expenses (in the range of 5–25% of conventional activated sludge treatment plants).

These low-cost sewage systems differ from traditional sewer designs as they prioritize transporting wastewater that is free of solids through the system. This approach makes them suitable for community management and allows for easy expansion and extension of existing systems. Additionally, they can connect satellite communities to centralized systems, making them adaptable and flexible for various settings. Such systems have also been successfully used in refugee camps to address wastewater management needs efficiently.

However, a limitation of these low-cost sewage systems is that they are not suitable for stormwater drainage. They are designed to handle wastewater and may not effectively manage stormwater runoff.

Furthermore, healthy ecosystems can provide cost-effective wastewater treatment services under certain conditions. Proper regulation of pollutant loads and types of contaminants in the effluent is necessary to ensure the ecosystem's effectiveness. Additionally, it is crucial not to exceed the pollution-carrying capacity of the ecosystem to maintain its ability to treat wastewater efficiently [22].

5.3 The Use of Wastewater as an Alternative Water Source

For centuries, the practice of using untreated or diluted wastewater for irrigation has been common. However, with growing urbanization and increased demand for water, many cities are now exploring alternative water sources, and reclaimed water has become an attractive option for sustainable water supplies to industries and municipalities. Reusing treated wastewater becomes economically viable when the point of reuse is close to the point of production.

Treating wastewater to meet quality standards for its intended use enhances the potential for cost recovery. Additionally, the competitiveness of wastewater use improves when freshwater prices reflect the true opportunity cost of using freshwater, and pollution charges adequately cover the expenses of removing pollutants from wastewater.

Strategic planning for the use of treated and partially treated wastewater for ecosystem services offers numerous benefits. It enhances resource efficiency by reducing freshwater withdrawals and promotes nutrient recycling. This approach also helps maintain healthier fisheries and aquatic ecosystems by minimizing water pollution and contributes to replenishing depleted aquifers [23].

5.4 Recovering Valuable By-Products: Extracting Resources from Wastewater

Wastewater has untapped potential as a valuable resource, particularly in energy generation and nutrient recovery. Integrating sludge/biosolids treatment processes into wastewater treatment plants can make them energy neutral or even net energy producers, reducing costs and carbon footprint while creating revenue opportunities through carbon credit programs. Advancements in technology allow for the recovery of nitrogen and phosphorus from wastewater and sewage sludge, offering a feasible and eco-friendly solution. Additionally, urine collection and utilization hold promise in ecological wastewater management as a sustainable source of essential nutrients for plant growth, especially as traditional phosphorus resources become scarce. Overall, these approaches contribute to the efficient and sustainable use of wastewater as a valuable resource [24].

6 Wastewater Generation and Treatment Around the World

The lack of available data on wastewater generation, collection, and treatment highlights a global issue: a significant portion of the world's wastewater is neither collected nor treated. Often, even when wastewater is collected, it is directly discharged into the environment without undergoing any treatment. Agricultural runoff, a type of wastewater, is seldom collected or treated, leading to a lack of measurement indicators for this category of wastewater. According to AQUASTAT, the database of the Food and Agriculture Organization of the United Nations (FAO), the annual freshwater withdrawals amount to 3,928 km^3. Out of this, approximately 44% (1,716 km^3) is consumed, primarily through evaporation from irrigated farmland in agriculture. The remaining 56% (2,212 km^3) is released back to the environment as wastewater, including municipal and industrial effluents and agricultural drainage water.

The level of industrial and municipal wastewater treatment at the country level often correlates with the income level of the country. High-income countries, on average, treat about 70% of the wastewater they generate, whereas upper-middle-income countries treat about 38%, and lower-middle-income countries treat around 28%. In low-income countries, the treatment of industrial and municipal wastewater is significantly lacking, with only about 8% of this wastewater receiving any treatment. This disparity in treatment rates indicates a need for improved wastewater management practices, particularly in less economically developed regions. Addressing the inadequate treatment and disposal of wastewater is crucial for preserving water quality and mitigating environmental pollution [19]. The lack of proper water and sanitation services in impoverished areas, particularly in slums, exacerbates the challenges related to wastewater management. These communities are directly exposed to untreated wastewater, leading to significant health risks. Globally, it is estimated that approximately 80% of wastewater is released into the environment without adequate treatment, compounding the issue.

There are notable disparities in wastewater treatment across different regions. In Europe, about 71% of municipal and industrial wastewater is treated, while the treatment rate is much lower in Latin American countries, at only 20%. The Middle East and North Africa (MENA) region treat an estimated 51% of municipal and industrial wastewater. In African countries, financial constraints hinder the development of proper wastewater treatment infrastructure, resulting in limited progress in wastewater management. Shockingly, 32 out of 48 countries in Sub-Saharan Africa have no data available on wastewater generation and treatment.

In regions with humid climates and high-income countries, such as North America, Northern Europe, and Japan, strict regulations on effluent discharge and greater public awareness of environmental quality drive wastewater treatment and its proper use or disposal.

Overall, addressing the disparities in wastewater management and promoting proper treatment methods are essential steps in improving public health and environmental conditions, particularly for marginalized communities facing direct exposure to untreated wastewater [19]. Indeed, the situation in high-income countries in drier regions, such as parts of North America, Australia, the Middle East, and southern Europe, differs from that in other regions. Due to increasing competition for water use between agriculture and other sectors in these water-stressed regions, treated wastewater is often utilized for irrigation to meet the demand for water resources.

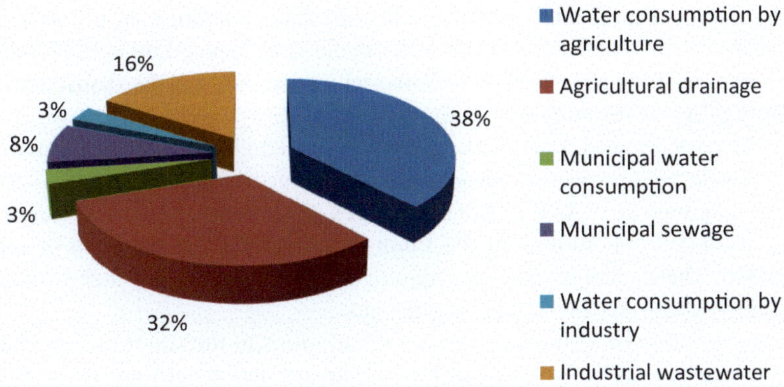

Fig. 1 Consumption and production of wastewater by the main water-consuming sectors

Insufficient treatment and discharge of wastewater strain existing facilities, leading to sub-optimal performance and compromising water quality. Untreated or undertreated wastewater is discharged into water bodies, negatively impacting downstream users and the environment. Improved wastewater management is crucial to safeguard water resources and community well-being.

To effectively address these challenges, proper operation, maintenance, and expansion of treatment facilities are essential. Additionally, stringent regulatory measures and efficient wastewater management practices are necessary to ensure that treated wastewater meets acceptable quality standards before being discharged into the environment. By optimizing wastewater treatment processes and improving the overall management of wastewater resources, it becomes possible to mitigate the impact on water quality and availability for downstream users in these water-scarce regions (Fig. 1).

7 Global Availability of Water Resources: Increasing Scarcity Each year

Water resources, including surface water and groundwater, are replenished through the ongoing process of the water cycle. This natural cycle involves the evaporation of water from oceans, lakes, and other water bodies, which then forms clouds. These clouds release precipitation in the form of rain or snow, which eventually falls back to the Earth's surface. Some of the precipitation directly replenishes surface water sources like rivers and lakes, while some percolates into the ground, recharging groundwater aquifers.

The water cycle is influenced by global and climatic forces, which cause fluctuations in precipitation and evaporation rates. These variations, in turn, affect the patterns of water runoff and the availability of water resources in different locations and over time. Additionally, the water cycle is influenced by natural and artificial storage, such as reservoirs and dams, which can impact the distribution and availability of water resources.

Over the past decades, observations have indicated changes in the dynamics of the water cycle, and projections from climate change scenarios suggest that these variations may intensify. As the climate changes, there can be shifts in precipitation patterns, leading to alterations in the timing, amount, and distribution of rainfall and snowfall. These changes can have significant implications for water availability, water quality, and water management practices.

The exacerbation of spatial and temporal variations in the water cycle dynamics due to climate change may result in more frequent and severe droughts in some regions, while other areas might experience increased rainfall and flooding. These changes can challenge water resource planning, agriculture, and other water-dependent sectors. Effective management and adaptation strategies are crucial to address the potential impacts of climate-induced changes in the water cycle and to ensure sustainable water resource utilization in the future [25]. The increasing gap between water supply and demand is exacerbating worldwide. Recent research reveals that approximately two-thirds of the global population currently resides in regions facing water scarcity for at least one month per year. Notably, about 50% of those affected by this level of water scarcity are located in China and India. Roughly 500 million people live in areas where water consumption is twice the local renewable water resources [26]. This is evident in parts of India, China, the Mediterranean region, the Middle East, Central Asia, the arid regions of sub-Saharan Africa, Australia, central and western South America, as well as central and western North America. Regions reliant on non-renewable resources like fossil groundwater, which is inherently unsustainable, are facing a decline and have become highly vulnerable, relying on water transfers from water-rich areas.

While floods and droughts are natural phenomena inherent in the dynamics of the water cycle, the frequency and severity of these events have changed significantly in many river basins worldwide due to a combination of climate change and human activities. Human-induced changes in land use, such as urbanization and river channelization, are altering the storage capacity of river basins, affecting both high and low flows, groundwater recharge, and runoff generation. Consequently, the frequency of water-related disasters is increasing.

Addressing these challenges necessitates comprehensive approaches, including efficient water use, conservation, infrastructure development, and proactive measures to adapt to changing climate patterns. Integrated water resource management and international cooperation for water-sharing arrangements are essential to ensure equitable access to water resources and reduce vulnerabilities in water-stressed regions.

Municipal wastewater is a composite of waste from domestic, industrial, commercial, and institutional activities within a specific human settlement or community. It comprises both municipal wastewater, which is the result of daily activities in the urban area, and urban runoff, which includes stormwater and rainwater runoff from streets and other surfaces.

The volume and composition of municipal wastewater are closely linked to the structure and operations of urban systems. As urbanization continues to evolve, it becomes essential to carefully analyze current and future patterns of urban development. Understanding these patterns is crucial for developing sustainable strategies for wastewater management in the years to come.

By taking into account the dynamics of urban growth and assessing the impacts of urbanization on wastewater generation, communities can implement more effective and environmentally friendly approaches to manage and treat wastewater. Emphasizing sustainable wastewater management practices is vital for safeguarding public health, preserving water resources, and minimizing the environmental impacts associated with urban development and population growth.

> As of my last update in September 2021, the percentage of the population served by different types of sanitation systems can vary significantly from country to country and even within regions of a country. Below are some common types of sanitation systems and their typical coverage:
> Improved Sanitation Facilities: This includes flush toilets connected to sewer systems or septic tanks, ventilated improved pit latrines, and composting toilets. The coverage of improved sanitation facilities can vary widely depending on the country's development level, urbanization, and infrastructure. In some developed countries, coverage can be close to 100%, while in developing countries, it may range from a few percent to over 50%.
> Unimproved Sanitation Facilities: This includes open defecation or using basic pit latrines without proper containment. In many developing countries, particularly in rural areas, unimproved sanitation facilities are still prevalent. The percentage of the population relying on unimproved sanitation can range from a few percent to over 50%, depending on the country.

Shared Sanitation Facilities: Some communities or households may share sanitation facilities, such as public toilets or community latrines. The percentage of the population relying on shared sanitation can also vary, but it tends to be more common in densely populated urban areas with limited access to private sanitation facilities.

It is important to note that the data on sanitation coverage is constantly changing due to population growth, urbanization, and efforts by governments and organizations to improve access to sanitation facilities. Therefore, for the most up-to-date and accurate information on the percentage of the population served by different types of sanitation systems in a specific country or region, it is best to refer to reliable sources like the World Health Organization (WHO) or the United Nations (UN) reports and databases.

8 Conclusion

In conclusion, the preservation and recovery of water resources' quality are crucial challenges for our society, both in the short term and for future generations. To address these challenges, various international and national initiatives have been established, such as the Water Framework Directive in Europe, the African Water Vision 2025, and commitments from the Grenelle Environment Forum in France. These initiatives aim to restore and protect water quality and ensure sustainable water management.

Water pollution, both anthropogenic and natural, poses significant threats to water resources. Anthropogenic pollution from agriculture, industries, and domestic activities contributes to the degradation of water quality. As a result, it is essential to identify and monitor the sources and vectors of pollution to implement appropriate measures for sustainable resource management.

Groundwater, despite being less directly exposed to pollution, can persistently retain contaminants and significantly impact surface water quality. The surface water-groundwater continuum is critical in protecting water resources, especially considering the vital role groundwater plays in providing base flow to rivers and supporting various human activities.

The generation and treatment of wastewater also play a significant role in the overall water management scenario. Globally, a considerable percentage of wastewater goes untreated, particularly in low and middle-income countries, leading to adverse impacts on human health, the environment, and economic activities. Implementing effective wastewater management strategies, such as prevention, collection, treatment, and reuse, is essential to alleviate water scarcity and pollution.

Moreover, sustainable wastewater management offers opportunities for resource recovery. The use of treated wastewater for irrigation and recycling valuable by-products can contribute to efficient water use and address increasing water scarcity.

Addressing the challenges of water resource preservation and recovery requires a comprehensive approach that combines sound regulatory frameworks, infrastructure development, public awareness, and cooperation among nations. By adopting sustainable practices and investing in effective wastewater management, we can safeguard water resources for current and future generations, ensuring a healthier and more sustainable environment for all.

References

1. 'Water Framework Directive'. https://environment.ec.europa.eu/topics/water/water-framework-directive_en. Accessed 06 Aug 2023
2. Donkor, S. M.: 'African Water Vision'
3. 'Ecological status of surface waters in Europe'. https://www.eea.europa.eu/ims/ecological-status-of-surface-waters. Accessed 06 Aug 2023
4. Water pollution from agriculture: a global review—Executive summary'

5. 'Groundwater Contamination—an overview | ScienceDirect Topics'. https://www.sciencedirect.com/topics/earth-and-planetary-sciences/groundwater-contamination. Accessed 06 Aug 2023
6. 'Water | Free Full-Text | Assessing the role of water resources protection practice for sustainable water resources management: a review. https://www.mdpi.com/2073-4441/14/19/3153. Accessed 06 Aug 2023
7. 'Water use and environmental pressures—European Environment Agency'. https://www.eea.europa.eu/themes/water/european-waters/water-use-and-environmental-pressures/water-use-and-environmental-pressures. Accessed 06 Aug 2023
8. 'Wastewater as a resource—May 2022—World | ReliefWeb', 18 Aug, 2022. https://reliefweb.int/report/world/wastewater-resource-may-2022. Accessed 06 Aug 2023
9. 'Water | Free Full-Text | Insights into the Domestic Wastewater Treatment (DWWT) Regimes: A Review'. https://www.mdpi.com/2073-4441/14/21/3542. Accessed 06 Aug 2023
10. Abdelfattah, A., et al.: Microalgae-based wastewater treatment: Mechanisms, challenges, recent advances, and future prospects. Environ. Sci. Ecotechnol. **13**, 100205 (2023). https://doi.org/10.1016/j.ese.2022.100205
11. Vardhan, K.H., Kumar, P.S., Panda, R.C.: A review on heavy metal pollution, toxicity and remedial measures: current trends and future perspectives. J. Mol. Liq. **290**, 111197 (2019). https://doi.org/10.1016/j.molliq.2019.111197
12. 'Sustainability | Free Full-Text | The potential of industrial symbiosis: case analysis and main drivers and barriers to its implementation. https://www.mdpi.com/2071-1050/11/24/7095. Accessed 06 Aug 2023
13. 'World's Demand for Food and Water: The Consequences of Climate Change | IntechOpen'. https://www.intechopen.com/chapters/66882. Accessed 06 Aug 2023
14. 'Agronomy | Free Full-Text | Towards sustainable application of wastewater in agriculture: a review on reusability and risk assessment'. https://www.mdpi.com/2073-4395/12/6/1397. Accessed 06 Aug 2023
15. 'UN-Water Annual Report 2015 | UN-Water'. https://www.unwater.org/publications/un-water-annual-report-2015. Accessed 06 Aug 2023
16. Onu, M.A., Ayeleru, O.O., Oboirien, B., Olubambi, P.A.: Challenges of wastewater generation and management in sub-Saharan Africa: a review. Environ. Chall. **11**, 100686 (2023). https://doi.org/10.1016/j.envc.2023.100686
17. Abd-Elhamid, H.F., Abd-Elmoneem, S.M., Abdelaal, G.M., Zeleňáková, M., Vranayova, Z., Abd-Elaty, I.: Investigating and managing the impact of using untreated wastewater for irrigation on the groundwater quality in arid and semi-arid regions. Int. J. Environ. Res. Public Health **18**(14), 7485 (2021). https://doi.org/10.3390/ijerph18147485
18. Hughes, J., Cowper-Heays, K., Olesson, E., Bell, R., Stroombergen, A.: Impacts and implications of climate change on wastewater systems: a New Zealand perspective. Clim. Risk Manag. **31**, 100262 (2021). https://doi.org/10.1016/j.crm.2020.100262
19. 'Global report on Sanitation and Wastewater Management in cities and human settlements | Department of Economic and Social Affairs'. https://sdgs.un.org/partnerships/global-report-sanitation-and-wastewater-management-cities-and-human-settlements. Accessed 06 Aug 2023
20. 'The Sources and Solutions: Wastewater | US EPA'. https://www.epa.gov/nutrientpollution/sources-and-solutions-wastewater. Accessed 06 Aug 2023
21. Azoulay, K., Bencheikh, I., Samghouli, N., Mabrouki, J., Moufti, A., El Hajjaji, S.: Modeling and Design of Water Treatment Processes by Biosorption Method Using JMP®11 Software. In: Azrour, M., Irshad, A., Chaganti, R., Eds., IoT and smart devices for sustainable environment, in EAI/Springer Innovations in Communication and Computing. Cham: Springer International Publishing, 2022, pp. 53–69. https://doi.org/10.1007/978-3-030-90083-0_5
22. Bencheikh, I., Azoulay, K., Ben Baaziz, M., Mabrouki, J.: Ecological risks related to the influence of different environmental parameters on the microplastics behavior. In: Mabrouki, J., Mourade, A., Irshad, A., Chaudhry, S. A., Eds., Advanced technology for smart environment and energy, environmental science and engineering. Cham: Springer International Publishing, 2023, pp. 117–128. https://doi.org/10.1007/978-3-031-25662-2_10

23. Karima, A.: Étude expérimentale et modélisation du procédé d'adsorption sur déchets de palmier : Application au traitement d'effluents textiles, université Mohammed V, faculté des sciences Rabat, 2021, Maroc.
24. Bencheikh, I. et al.: Mathematical and statistical study for the wastewater adsorbent regeneration using the central composite design. In: Azrour, M., Irshad, A., Chaganti, R., Eds., IoT and Smart Devices for Sustainable Environment, in EAI/Springer Innovations in Communication and Computing. Cham: Springer International Publishing, 2022, pp. 71–83. https://doi.org/10.1007/978-3-030-90083-0_6
25. 'AR5 Climate Change 2013: The Physical Science Basis — IPCC'. https://www.ipcc.ch/report/ar5/wg1/. Accessed 06 Aug 2023
26. Mekonnen, M.M., Hoekstra, A.Y.: Four billion people facing severe water scarcity. Sci. Adv. 2, e1500323 (2016).—References—Scientific Research Publishing'. https://www.scirp.org/(S(lz5mqp453edsnp55rrgjct55.))/reference/referencespapers.aspx?referenceid=2900611. Accessed 06 Aug 2023

Application of Machine Learning Algorithms for Air Quality Prediction

Souad Nasrdine, Jamal Mabrouki, Najlaa Ben-Lhachemi, and Miloudia Slaoui

Abstract Since becoming one of the leading causes of respiratory illnesses, heart problems and even death, air pollution has been a major concern globally. Various techniques and technologies have been created to monitor and manage air pollution, but deep learning techniques have shown great promise in their ability to accurately predict pollutant concentrations. Machine learning algorithms have been studied and developed recently with the aim of predicting concentrations of airborne contaminants and implementing preventive actions to protect health. The main objective of this article is to perform an analysis of the results of the concentration of CO in the air with R language using the ARIMA model to predict the air quality in the city of Rabat, Morocco.

Keywords Pollution · Machine learning algorithms · Pollutants · R language · ARIMA

1 Introduction

Air pollution is one of the most pressing sustainable development issues facing countries around the world [1]. According to the Sustainable Development Goals, air pollution is considered a major concern [2]. The World Health Organization has also estimated that air pollution is one of the top environmental health issues affecting people around the world, causing 4.2 million premature deaths in 2016 alone affects both urban and rural areas [3]. In addition, air pollution is proven to be

S. Nasrdine (✉) · N. Ben-Lhachemi · M. Slaoui
Energy, Materials and Sustainable Development (EMDD) Laboratory—Higher School of Technology—SALE, Center of Water, Natural Resources, Environment and Sustainable Development (CERN2D), University Mohammed V in RABAT, 10100 Rabat, Morocco
e-mail: souadnasrdine@gmail.com

J. Mabrouki
Laboratory of Spectroscopy, Molecular Modeling, Materials, Nanomaterials, Water and Environment, Faculty of Science CERNE2D, Mohammed V University in Rabat, Rabat, Morocco

J. Mabrouki and M. Azrour (eds.), *Integrated Solutions for Smart and Sustainable Environmental Conservation*, Studies in Systems, Decision and Control 527,
https://doi.org/10.1007/978-3-031-55787-3_7

a significant risk factor for disease burden, including cancer and asthma [4]. Morocco is also among these countries. It has recently experienced an increase in the rate of pollution, more precisely in the large cities which are experiencing an expansion of industrial activities such as the city of Rabat. Hence the importance of monitoring and improving air quality in cities [5]. One of the main challenges to achieve this goal is the analysis of the large amount of air quality data to determine the spatio-temporal characteristics and causes of pollution [6]. This is where machine learning comes into play as a new and innovative technology to predict and analyze air pollution [7]. These last years, many machine learning strategies have been proposed to address the issue of air quality forecasts in urban areas. These strategies employ data-driven methods that rely on sensors monitoring various air pollutants, weather information, and other relevant factors that influence air quality. This chapter will focus on the study of air quality in the city of Rabat using machine learning with R language, for better prediction the ARIMA model was chosen as a data analysis tool [8].

2 Machine Learning History

The goal of artificial intelligence research in the field of machine learning is to create statistical models and algorithms that allow machines to learn from data rather than being explicitly programmed to perform a particular task. The first applications of machine learning in the study of air quality appeared in the 2000 and focused on the study of various methods, including artificial neural networks; Wu et al. [9] focused on the use of artificial neural network (ANN) models to predict the PM10 pollution index in the urban area of Wuhan, China. Support vector machines (SVM) used by Stephane Canu and Alain Rakotomamonjy [10] for forecasting ozone peaks and air pollution. These methods have been used to predict air pollution levels based on historical data. In 2010 Researchers began to integrate various data sources to improve air quality prediction due to the expansion of data sources now available [11]. These include the combination of meteorological, geographical, socio-economic data and air quality sensors. Machine learning was used to analyze these numerous datasets and extract sophisticated patterns. Advances in machine learning have led to the development of more advanced models that have been used in air quality research. For example, deep neural networks, such as convolutional neural networks (CNN) [12, 13] and recurrent neural networks (RNN) [14, 15], have been used to capture temporal and spatial patterns in data on air quality. Research continued to progress for years in 2018 Shawabkeh et al. [16] determined benzene concentration using an artificial neural network (ANN) and a support vector machine, but did not plan not the error with precision between the actual value and the predicted value and its association with CO can be determined using the correlation method. In 2019 Baran [17] proposed air quality prediction using extreme learning machine (ELM). In 2020 Wang et al. [18] used gas recurrent neural network to predict air quality level and compare it with MLP and SVR. The gas recurrent neural network performs better due to less invariant sensor drift, however, the gas recurrent network is more sensitive to atmospheric variability and humidity. In 2023 Wei and Runsu [19] used historical air quality data from monitoring stations and integrated different environmental variables such as temperature, humidity, wind speed, etc., to form a prediction model.

3 Materials and Methods

3.1 Study Zone

In our study, we base ourselves on air quality data for the city of Rabat, which is located on the northwestern part of Morocco, with a view of the Atlantic Ocean, also undergoes the influence of these two centers of action on its behavior and climatic conditions. In addition to its geographical position, the region of Rabat is subject to oceanic influences, thus generating a temperate climate exposed to disturbances from the North and North-West of the country. Latitude: 34°00′47″ North, Longitude: 6°49′57″ West.

3.2 The Measurement Method Used

The monitoring station was chosen as the measurement method for our project: This station has specialized devices and sensors that are used to measure and record various parameters related to air quality. When she collects information on air pollution, she is usually found in urban, industrial or rural areas.

It has a number of advantages, one of which is that it was designed to operate continuously, allowing data to be collected over an extended period. This makes it a popular choice for measuring air pollutants. This helps to understand daily, seasonal and annual fluctuations in air pollution levels and to get a more complete picture of those levels.

3.3 R Language

R is an open-source programming language widely used for data analysis, statistical modeling, and data visualization. It is used in many fields. It is capable of processing large data sets. Although air quality data is sometimes very large and complicated, R can process and analyze it easily. It can be used by researchers to clean and prepare high-quality air data for analysis, as well as to conduct advanced statistical studies [20].

R is useful for simulating the effects of air pollutants on human and environmental health. The development of models that predict air pollution levels in various regions as well as the assessment of the effects of these levels on human and environmental health are both possible through the use of R. These models can help local authorities to make clear decisions about air quality [21].

3.4 The ARIMA Model

The choice of the ARIMA model (auto regressive integrated moving average) for our study is based on the fact that our data is formed as a time series, it is a statistical model designed for the analysis and prediction of time series data. It is composed of three parts: auto regression (AR), moving average (MA) and integration (I). it has several advantages namely the use of trends, seasons and cycles to simulate complex cyclical series, it takes into account the effects of autocorrelation and heteroscedasticity in the data and it provides accurate predictions for future values of the time series [22]. Will be some time decay and that the current value of the series will depend on its past values. This translates mathematically by:

$$\forall t : Xt = \sum_{i=1}^{p} \alpha_i X_{t-1} + \varepsilon_t \tag{1}$$

With ε the error and (α1, …, αp) reals.

The moving average (MA) method predicts future values using the historical errors of a time series. It assumes that the current value of the series depends on the past errors of the series, with some time decay. Mathematically this means that:

$$X_t = \varepsilon_t + \sum_{i=1}^{\rho} \alpha_i \varepsilon_{t-1} \tag{2}$$

The integration (I): is a technique that makes it possible to make a series chronologically stationary by differentiating it. To model and predict time series, stationarity is necessary because it allows to assume that the statistical properties of the series remain constant over time.

$$X_t = \beta_t + \alpha + \varepsilon_t \tag{3}$$

$$X_t - X_{t-1} = \beta + \varepsilon_t - \varepsilon_{t-1} \tag{4}$$

4 Result and Discussion

Carbon monoxide CO is a colorless and toxic gas, it is considered among the most dangerous pollutants for human health. For our study, we used the results of CO concentration in the air for 3 years: 2017, 2018, 2019 which are presented in the Fig. 1 below.

We can observe from these graphs that the concentrations of CO in the months of January until June are almost the same while in July and August we notice the difference and from September the concentrations converge towards the same values.

4.1 ARIMA Model

We used the ARIMA model to analyze and predict the data in the figures above and we obtained the following Table 1.

Predicted air quality values are shown in the "Point Forecast" column at all times.

The "Lo 80" and "Hi 80" columns indicate the lower and upper limits of the 80% confidence interval, respectively. This means that there is an 80% chance that the actual air quality values will fall within these ranges.

In addition, the "Lo 95" and "Hi 95" columns indicate the lower and upper limits of the 95% confidence interval. This means that there is a 95% chance that the actual air quality values will fall within these ranges.

We have the predetermined values as well as the lower and upper limits of the confidence interval for each predicted period. These confidence intervals give us an idea of the accuracy of your predictions. For example, the actual air quality values for the pollutant CO for the first time (corresponding to a value of 4.000000) have

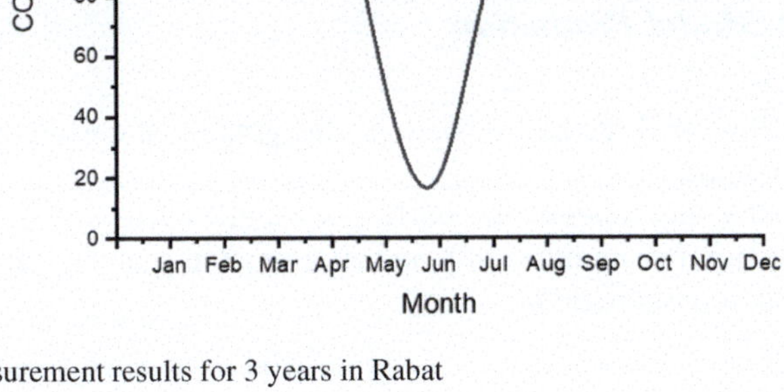

Fig. 1 CO measurement results for 3 years in Rabat

Table 1 Prediction results with ARIMA

	Point forecast	Lo 80	Hi 80	Lo 95	Hi 95
4.000000	135.00000	103.93266	166.06734	87.486603	182.51340
4.041667	138.51764	107.45030	169.58498	91.004239	186.03103
4.083333	142.09803	111.03069	173.16537	94.584633	189.61143
4.125000	147.09803	116.03069	178.16537	99.584633	194.61143
4.166667	148.54901	117.48167	179.61636	101.035618	196.06241
4.208333	150.83921	119.77187	181.90655	103.325815	198.35261
4.250000	148.45099	117.38364	179.51833	100.937589	195.96438
4.291667	146.74118	115.67384	177.80852	99.227786	194.25458
4.333333	141.03138	109.96404	172.09872	93.517983	188.54478
4.375000	132.90197	101.83463	163.96931	85.388574	180.41537
4.416667	122.90197	91.83463	153.96931	75.388574	170.41537
4.458333	112.58039	81.51305	143.64773	65.066997	160.09379
4.500000	106.16079	75.09345	137.22813	58.647391	153.67418
4.541667	100.32158	69.25424	131.38892	52.808180	147.83497
4.583333	101.16079	70.09345	132.22813	53.647391	148.67418
4.625000	49.13943	18.07209	80.20677	1.626033	96.65283
4.666667	102.58039	71.51305	133.64773	55.066997	150.09379
4.708333	105.83921	74.77187	136.90655	58.325815	153.35261
4.750000	107.00000	75.93266	138.06734	59.486603	154.51340
4.791667	110.83921	79.77187	141.90655	63.325815	158.35261
4.833333	115.00000	83.93266	146.06734	67.486603	162.51340
4.875000	120.00000	88.93266	151.06734	72.486603	167.51340
4.916667	136.38044	105.31310	167.44778	88.867046	183.89384
4.958333	142.02749	110.96015	173.09483	94.514091	189.54088

an 80% chance of being between 103.93266 and 166.06734, and 95% probability of being between 87.486603 and 182.51340.

4.2 ARIMA Model Prediction

We used the function plot (predictions), to get the graph that corresponds to the CO values in ppb (Fig. 2).

It can be seen that the values evolve fairly consistently each year. With a minimum value during the summer months (July and August) and a maximum value during the month of March (cold periods).

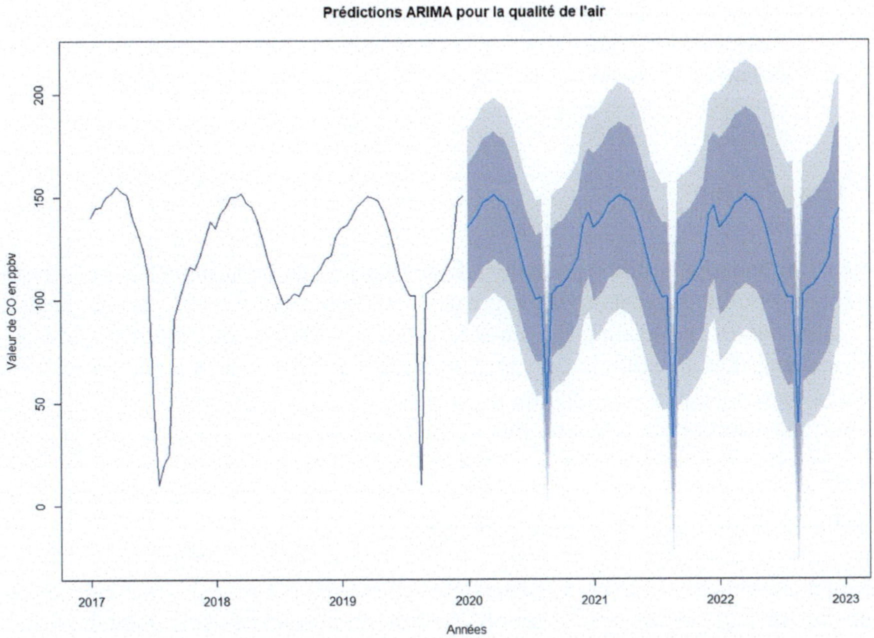

Fig. 2 Prediction of the ARIMA model for CO

The predicted and graphed values from the fourth year (2020) follow a similar pattern to previously recorded years (a peak during cold seasons and a decline during summer seasons).

This may indicate that the model has successfully captured trends and seasonal themes of our time series.

Since CO is a toxic gas that can negatively impact human health, the results of our analysis based on CO measurements in the air have important implications for human health. Long-term exposure to high levels of CO can lead to respiratory problems, cardiovascular disease, and even death. It is therefore essential to take measures to reduce the levels of CO in the air in order to protect both the environment and the health of the inhabitants of Rabat; CO is an air pollutant that contributes to air pollution. Our results highlight a significant source of CO that can result from a variety of human activities, including fossil fuel combustion, vehicle emissions, and industrial processes. These CO emissions affect air quality, which contributes to smog and other environmental problems. In order to preserve air quality and safeguard the environment, it is essential to reduce these CO emissions.

In conclusion, our results highlight the significant effects of CO measurement on human health, the environment, and air pollution control policies. They underline the need for immediate action to reduce CO levels in the air, safeguard people's health and preserve the environment. Air pollution policy management needs to be strengthened, and government organizations and policy makers need to implement stricter air quality standards and encourage the use of environmentally friendly technologies.

5 Conclusion

Around six and a half million premature deaths worldwide are attributed to indoor and outdoor air pollution, making it the greatest environmental threat to human health. It is also one of the leading preventable causes of disease and death worldwide. Recognizing that air pollution disproportionately affects women, children and the elderly, particularly in low-income populations who are exposed to high levels of indoor and outdoor air pollution due to cooking, wood and kerosene heating and other household practices. Due to its spread over such great distances and the lack of a firm response to address it, it is also a significant global problem. According to estimates, given the high costs of air pollution to society, including the negative effects on the economy, labor productivity, health care costs and tourism, as well as the financial benefits of investing in the fight against air pollution, it is recognized that there are financial arguments for action and that solutions exist. As a rental treatment of air pollution, poor air quality is a sustainability issue for all nations, especially in urban areas and cities where pollution levels exceed the upper limits set by the guidelines on air quality.

Some air pollutants, such as black carbon, methane, and surface ozone, are also short-lived climate pollutants that contribute significantly to air pollution-related deaths, cultural impacts, and food security. Reducing these pollutants will have positive effects on the climate.

In our study we explored the use of machine learning, more specifically the ARIMA model, for air quality prediction. Using R software, we analyzed CO measurement data for 3 years and developed an ARIMA model to make accurate predictions. The ability of the ARIMA model to accurately capture trends, seasonal themes and fluctuations in air quality has been demonstrated. The usefulness of the model for air quality management is evidenced by the fact that the predictions it produced agree with actual measurements. It is important to point out that this approach also has certain limitations. As forecasts are based on currently available historical data, unexpected events or significant changes may not be fully accounted for. Furthermore, the accuracy of the forecasts depends on the quality of the input data and the efficiency of the ARIMA model itself.

References

1. Bai, L., Wang, J., Ma, X., Lu, H.: Air pollution forecasts: an overview. Int. J. Environ. Res. Public Health **15**, 780 (2018)
2. Curtis, L., Rea, W., Smith-Willis, P., Fenyves, E., Pan, Y.: Adverse health effects of outdoor air pollutants. Environ. Int. **32**, 815–830 (2006)
3. Hannan, E.J., Quinn, B.G.: The determination of the order of an autoregression. J. R. Stat. Soc. B **41**, 190–195 (1979)
4. Bellinger, C., Mohomed Jabbar, M.S., Zaiane, O., Osornio-Vargas, A.: A systematic review of data mining and machine learning for air pollution epidemiology. BMC Public Health **17**, 907 (2017)
5. Elminir, H.K.: Dependence of urban air pollutants on meteorology. Sci. Total. Environ. **350**, 225–237 (2005)
6. Zhou, Q., Jiang, H., Wang, J., Zhou, J.: A hybrid model for PM2.5 forecasting based on ensemble empirical mode decomposition and a general regression neural network. Sci. Total Environ. **496**, 264–274 (2014)
7. Boudhar, A., Rhinane, H., Zarhloule, Y.: Trends in temperature and precipitation extremes over Morocco. Theoret. Appl. Climatol. **126**(3–4), 691–705 (2016)
8. Douaik, A., Radoane, M., Rhinane, H.: Changes in temperature and rainfall extremes over Rabat, Morocco. Theoret. Appl. Climatol. **128**(3–4), 791–803 (2017)
9. Wu, S., Feng, Q., Du, Y., Li, X.: Artificial neural network models for daily PM10 air pollution index prediction in the Urban Area of Wuhan, China. Environ. Eng. Sci. **28**(5), 2011. https://doi.org/10.1089/ees.2010.0219
10. Canu, S., Rakotomamonjy, A.: Ozone peak and pollution forecasting using support vectors (2000)
11. Dragomir, E.G.: Air quality index prediction using K-nearest neighbor technique. Bull. PG Univ. Ploiesti, Ser. Math. Inform. Phys. LXII **1**, 103–108 (2010)
12. Wang, J., Li, X., Jin, L., et al.: An air quality index prediction model based on CNN-ILSTM. Sci. Rep. **12**, 8373 (2022). https://doi.org/10.1038/s41598-022-12355-6
13. Chauhan, R., Kaur, H., Alankar, B.: Air quality forecast using convolutional neural network for sustainable development in urban environments. Sustain. Cities Soc. **75**, 103239 (2021). ISSN 2210-6707. https://doi.org/10.1016/j.scs.2021.103239
14. Athira, V., Geetha, P., Vinayakumar, R., Soman, K.P.: DeepAirNet: applying recurrent networks for air quality prediction. Procedia Comput. Sci. **132**, 1394–1403 (2018). ISSN 1877–0509. https://doi.org/10.1016/j.procs.2018.05.068
15. Cassano, F., Casale, A., Regina, P., Spadafina, L., Sekulic, P.: A recurrent neural network approach to improve the air quality index prediction. In: Novais, P., Lloret, J., Chamoso, P., Carneiro, D., Navarro, E., Omatu, S. (eds) Ambient Intelligence—Software and Applications—10th International Symposium on Ambient Intelligence. ISAmI 2019. Advances in Intelligent Systems and Computing, vol. 1006. Springer, Cham. https://doi.org/10.1007/978-3-030-24097-4_5
16. Shawabkeh, F. A., Rodan, A., Salem, M. : Modèle de surveillance de la pollution atmosphérique par le benzène utilisant ANN et SVM. IEEE (2018)
17. Baran, B.: Prédiction de l'indice de qualité de l'air par Extreme Learning Machines. IEEE (2019)
18. Wang, S., Hu, Y., Marco, J.B.S., Liu, S.-C.: Prédiction de la concentration de gaz à l'aide de réseaux de neurones récurrents fermés. IEEE (2020)
19. Fang, W., Zhu, R., Lin, J.C.: An air quality prediction model based on improved Vanilla LSTM with multichannel input and multiroute output. Exp. Syst. Appl. **211**, 118422 (2023). ISSN 0957-4174. https://doi.org/10.1016/j.eswa.2022.118422

20. Ihaka, R., Gentleman, R.: R: a language for data analysis and graphics. J. Comput. Graph. Stat. **5**(3), 299–314 (1996)
21. Lyazrhi, F.: Une introduction au langage R. Ecole Nationale Vétérinaire, **23** (2005)
22. Kumar, U., Jain, V.K.: ARIMA forecasting of ambient air pollutants (O_3, NO, NO_2 and CO). Stoch. Environ. Res. Risk Assess. **24**, 751–760 (2010)

Comparative Study of Spectrograms and Scalograms for Fetal Electrocardiogram Analysis in Healthcare: Unveiling the Trade-Offs Between Time and Frequency Resolution

Said Ziani and Achmad Rizal

Abstract This paper presents a comparative analysis of spectrograms and scalograms for ECG signal analysis, focusing on the trade-offs between time and frequency resolution. Previous studies have explored these techniques individually, but a systematic comparison considering various window types, wavelet families, and parameter settings is lacking. Our research fills this gap by conducting a comprehensive investigation to unveil the strengths and limitations of spectrograms and scalograms in capturing the complex features of mixed fetal and maternal ECG signals. Through extensive experimentation, we demonstrate that the choice of wavelet family significantly impacts the analysis results. Among the wavelets examined, the Symlet wavelet emerges as a powerful tool for optimal analysis of fetal and maternal activities. We highlight its ability to capture intricate details and patterns, showcasing its superior performance compared to other wavelet families.

Keywords Electrocardiogram · Spectrograms · Scalograms · Time-frequency Resolution

1 Introduction

Electrocardiogram (ECG) signals provide crucial information about the electrical activity of the heart and are extensively used in the diagnosis and monitoring of cardiac disorders. Analyzing ECG signals accurately is essential for detecting abnormalities, assessing the overall cardiac health, and guiding appropriate medical interventions. In recent years, time-frequency analysis techniques have gained significant popularity for analyzing non-stationary signals like ECG, as they provide

S. Ziani (✉)
ENSAM, Mohammed V University in Rabat, Rabat, Morocco
e-mail: s.ziani@um5r.ac.ma

A. Rizal
Telkom University, School of Electrical Engineering Telekomunikasi no.1, Bandung, Indonesia
e-mail: achmadrizal@telkomuniversity.ac.id

J. Mabrouki and M. Azrour (eds.), *Integrated Solutions for Smart and Sustainable Environmental Conservation*, Studies in Systems, Decision and Control 527,
https://doi.org/10.1007/978-3-031-55787-3_8

a comprehensive representation of signal dynamics in both the time and frequency domains. Two widely used time-frequency analysis tools for ECG signal analysis are spectrograms and scalograms. Spectrograms, based on the Short-Time Fourier Transform (STFT) [1], offer a detailed representation of signal energy distribution over time and frequency by dividing the signal into smaller segments and computing the Fourier Transform for each segment. On the other hand, scalograms, based on the Continuous Wavelet Transform (CWT) [2, 3], provide a multi-resolution analysis by convolving the signal with a family of wavelets at different scales, capturing both high and low-frequency components with enhanced time localization. The choice between spectrograms and scalograms depends on the specific requirements of the analysis, considering the trade-offs between time and frequency resolution. Spectrograms excel in capturing frequency details accurately but suffer from limited time localization, which may hinder the analysis of transient features in ECG signals. On the other hand, scalograms offer superior time localization and facilitate the precise identification of temporal characteristics, but they may have limited frequency resolution, particularly for high-frequency components. In order to guide researchers and academics in selecting the appropriate time-frequency representation technique for ECG signal analysis, this paper presents a comparative analysis of spectrograms and scalograms. The aim is to provide insights into the strengths and limitations of each method and to determine the optimal parameters for their utilization. The analysis encompasses evaluating various performance metrics, including time-frequency resolution, signal-to-noise ratio, and computational complexity. Furthermore, a diverse dataset of ECG signals representing different cardiac conditions is utilized for experimentation. This research builds upon the existing body of knowledge in the field of ECG signal processing and contributes to the understanding of the benefits and drawbacks associated with spectrograms and scalograms. The findings presented in this paper will assist researchers and practitioners in making informed decisions regarding the selection of the most suitable analysis tool based on their specific analysis requirements. By providing a comprehensive comparison of these mathematical microscopes, this work aims to enhance the accuracy and effectiveness of ECG signal analysis, ultimately leading to improved diagnosis and treatment of cardiac disorders.

2 Theoretical Framework and Research Methodology

2.1 Theoretical Framework

The Short-Time Fourier Transform (STFT) and Continuous Wavelet Transform (CWT) are two commonly used signal processing techniques for analyzing time-varying signals. Here's a brief explanation of their theoretical frameworks, along with the mathematical equations involved:

2.1.1 Short-Time Fourier Transform (STFT)

The STFT provides a time-frequency representation of a signal by decomposing it into short overlapping segments and computing the Fourier Transform on each segment. The STFT can reveal how the frequency content of a signal changes over time. Mathematically, given a signal x(t), the STFT is calculated as follows:

$$STFT(t, f) = \int [x(\omega) * w(t - \omega)]exp(-j2\pi\omega)d\omega \tag{1}$$

where t represents time, f represents frequency, ω represents a time-shift parameter, w(t-ω) is a window function applied to each segment, and $*$ denotes the convolution operation. The STFT output is a 2D matrix, where the x-axis represents time, the y-axis represents frequency, and the magnitude or phase of each element represents the signal's spectral content at a specific time and frequency.

2.1.2 Continuous Wavelet Transform (CWT)

The CWT analyzes signals in both time and frequency domains simultaneously. It uses a family of wavelet functions that are scaled and shifted to match different frequencies and positions in the signal, providing a time-frequency representation that adapts to the signal's local characteristics. Mathematically, given a signal x(t), the CWT is calculated as follows:

$$CWT(s, r) = \int [x(t) * \psi^*((t - r)/s)]dt, \tag{2}$$

where a represents the scale parameter, b represents the translation parameter, x(t) is the input signal, ψ(t) is the analyzing wavelet function, and $*$ denotes the convolution operation. The CWT output is a 2D matrix, where the x-axis represents time, the y-axis represents scale, and the magnitude or phase of each element represents the signal's similarity to the wavelet at a particular time and scale.

Both the STFT and CWT provide valuable insights into the time-frequency characteristics of signals, enabling various applications in fields such as audio processing, image analysis, and biomedical signal analysis.

2.2 *Research Methodology*

Our objective is to provide academics and researchers in the field of biomedical signal processing, specifically ECG signals, with essential parameters for conducting cost-effective time-frequency and time-scale analysis. This analysis aims to characterize the signals for diagnostic purposes.

When discussing the STFT, the major challenge lies in determining the optimal window nature, as well as the values of parameters such as nftt, hopsiese, overlap, and window length that allow for a proper representation of the signal. On the other hand, when addressing the CWT, the main concern is to determine the appropriate wavelet nature and scale range capable of highlighting the spectral content of the signal.

In this paper, we focused on an ECG signal captured by appropriately placed abdominal sensors on a pregnant woman's abdomen. Essentially, the measured ECG is a combination of the fetal ECG (FECG), related to the fetal heart's electrical activity, and the maternal ECG (MECG), related to the mother's heart electrical activity. Each of these signals is characterized by the presence of multiple waveforms, namely P, QRS, T, and U waves. However, from a medical perspective, the QRS complex is of particular interest for ECG signal analysis. Its absence or irregularity is an important pathological indicator that requires careful consideration.

We denote QRSF as the QRS complex of the fetus and QRSM as the QRS complex of the mother. Our goal is to determine how to judiciously apply both the STFT and CWT tools to effectively diagnose the ECG signal of a pregnant woman, while also highlighting the added value of the CWT compared to the STFT.

The proposed algorithm in Fig. 1 illustrates the methodology employed to achieve the aforementioned objectives.

3 Results and Discussion

3.1 Observed Signal

The performance of the algorithm is assessed using open source datasets, specifically the Daisy dataset, which comprises recordings of 5 abdominal channels and 3 thoracic channels, all with a sampling rate of 250 Hz [4] (Fig. 2).

3.2 Results and Discussion

3.2.1 STFT Findings

In this study, we present Fig. 3, which depict the 2D spectrograms of the filtered input signal using the four main window functions: Hanning, Hamming, Blackman, and Rectangular (Rectwin). These figures highlight the significance of the Blackman window in characterizing the QRSF and QRSM complexes related to the fetus and mother compared to the other mentioned windows [5]. To determine the optimal parameters for utilizing the Blackman window, we developed an algorithm that explores different window lengths and hop sizes. Figures 4, 5, 6 and 7 demonstrate the

Fig. 1 Algorithm

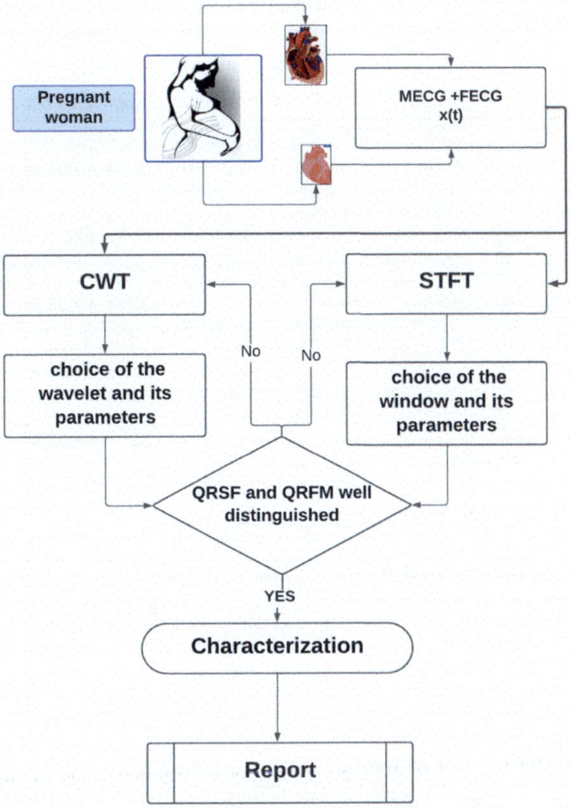

remarkable results obtained in 3D for the following cases: window-lengths = [256, 512, 1024, 2048]; hop-sizes = [128, 256, 512, 1024]; window-types = ["hanning", "hamming", "blackman", "rectwin"].

The key finding highlighted in this paper is the compromise proposed, known as the "Ziani compromise," which involves using the Blackman window with a window length of 256 and a hop size of 128. This configuration enables an optimal analysis of an ECG signal that comprises a mixture of fetal and maternal activities.

The results demonstrate that the Blackman window exhibits superior performance in capturing the QRSF and QRSM complexes, which are crucial in distinguishing the fetal and maternal components of the ECG signal. By carefully selecting the window length and hop size, we can achieve an optimal balance between time and frequency resolution, enhancing the accuracy of fetal and maternal activity characterization. The proposed methodology and findings have significant implications in biomedical literature. The "Ziani compromise" with the Blackman window and specific parameter settings provides an effective approach for analyzing ECG signals with mixed fetal and maternal activities. This research contributes to the advancement of signal processing techniques in the field of fetal monitoring, facilitating improved diagnosis and monitoring of fetal well-being during pregnancy as shown in Fig. 8.

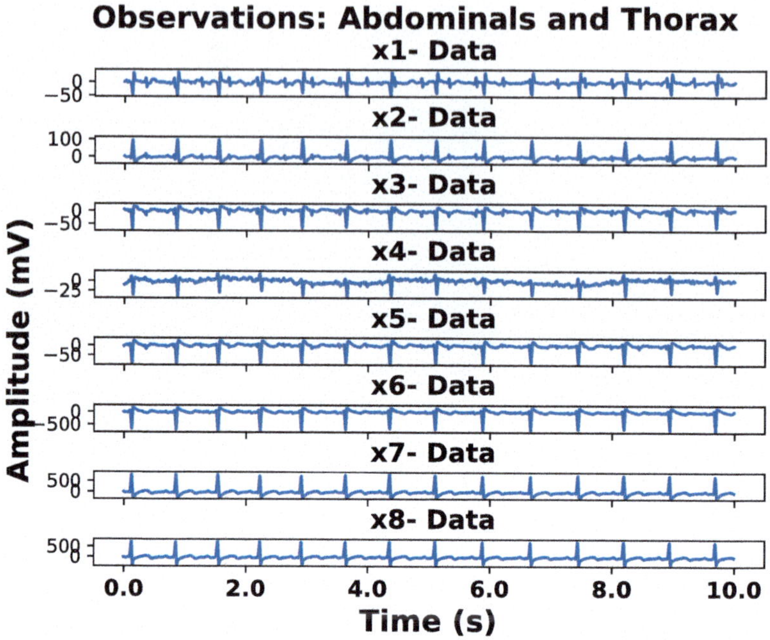

Fig. 2 Observed signal

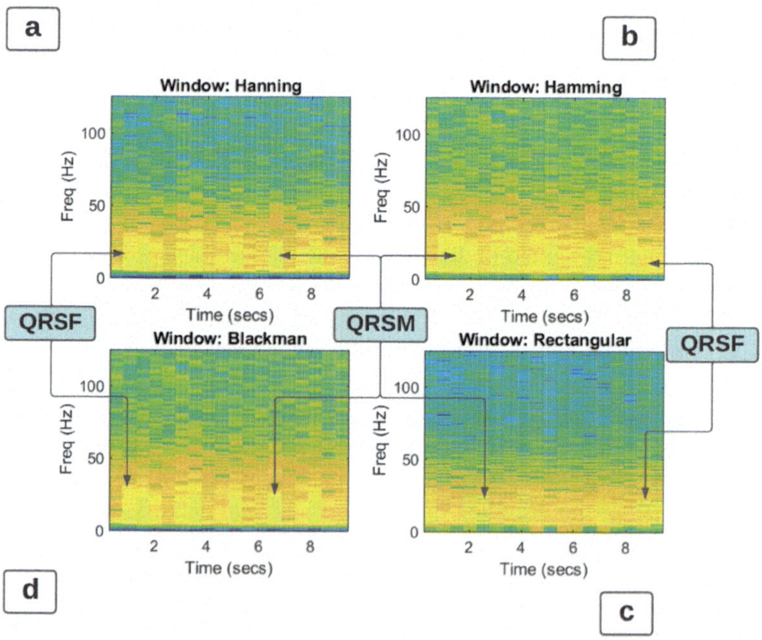

Fig. 3 Spectrogram by various window

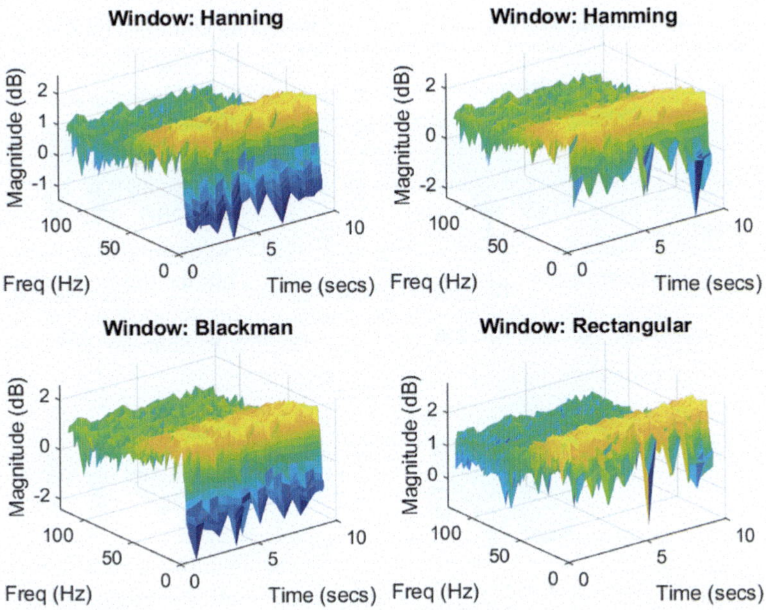

Fig. 4 Observed signal

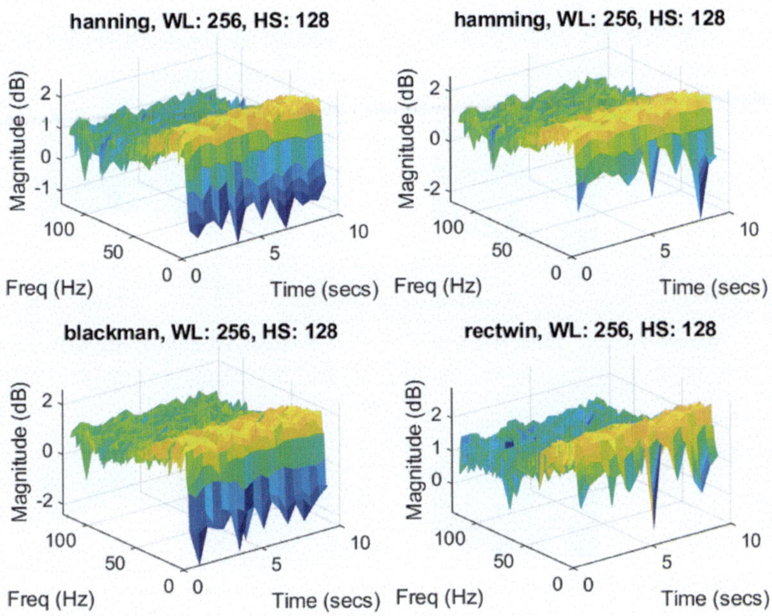

Fig. 5 Observed signal

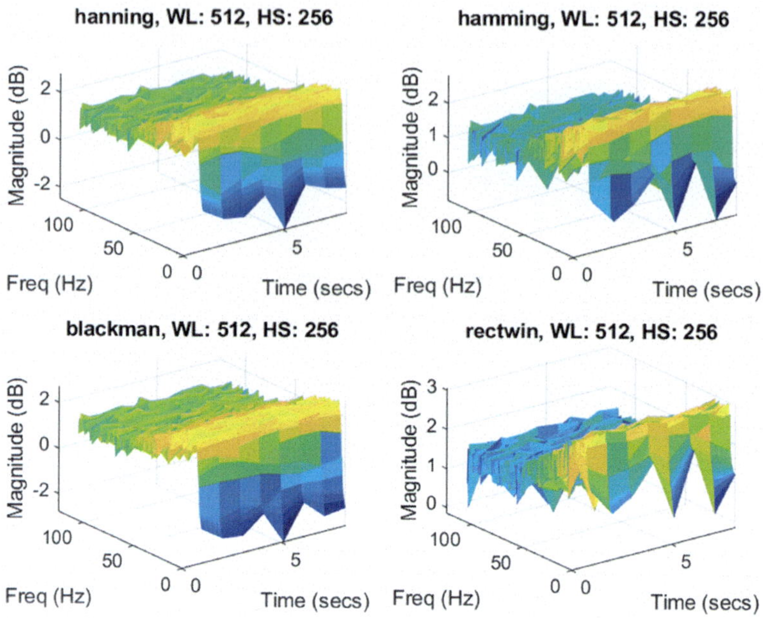

Fig. 6 Spectrogram by various window

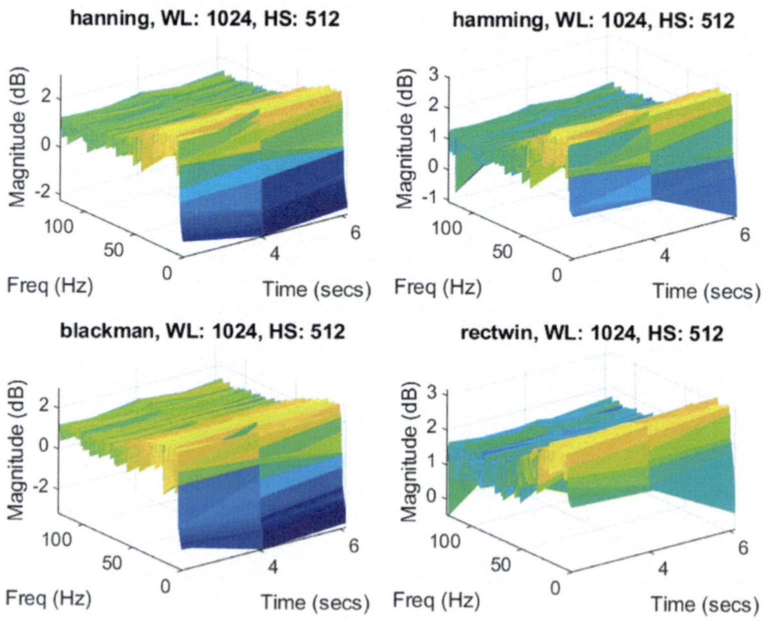

Fig. 7 Observed signal

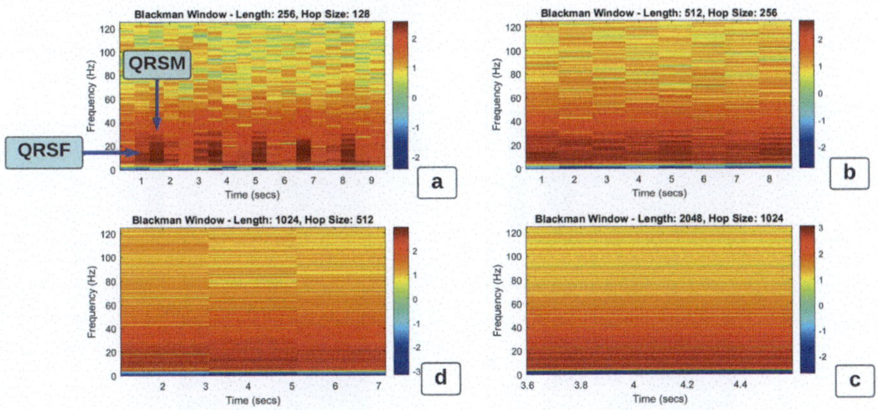

Fig. 8 Observed signal

3.2.2 CWT Fundings

A major drawback of the STFT tool lies in its limited spatial-temporal localization, which has historically been highlighted by the Schrödinger inequality. The STFT utilizes fixed window lengths and hop sizes, resulting in a trade-off between time and frequency resolution. This trade-off inherently leads to poor localization capabilities as the STFT cannot adapt to variations in signal characteristics across different time and frequency regions [6, 7].

However, the Continuous Wavelet Transform (CWT) provides a solution to this limitation. Unlike the STFT, the CWT employs a variable-width window known as the wavelet, which can be scaled and shifted to capture signal features at different scales and positions. This inherent flexibility allows for improved spatial-temporal localization, enabling precise identification and analysis of transient events and localized frequency components.

By utilizing the CWT, researchers and practitioners can overcome the challenges posed by the STFT's spatial-temporal localization. The CWT offers a more comprehensive representation of signals, delivering enhanced resolution and accuracy in both the time and frequency domains. Its adaptability to signal characteristics makes it a valuable tool in various applications, including biomedical signal analysis, image processing, and time-frequency analysis. Figures 9, 10, and 11 showcase the spectacular results obtained using different wavelets for analyzing a mixed fetal and maternal ECG signal. The figures display the 3D representations of the spectrograms generated by applying various wavelets to the signal. Specifically, the following wavelets were evaluated: Morlet, Daubechies, Symlet, Haar, Coiflet, Meyer, Mexican Hat (or Ricker wavelet), Shannon, BiorSplines, Gabor, and amor.

Among these wavelets, the noteworthy finding in this study is the superior performance of the Symlet wavelet. Figure 10, representing the spectrogram obtained with the Symlet wavelet, exhibits optimal analysis capabilities for extracting meaningful

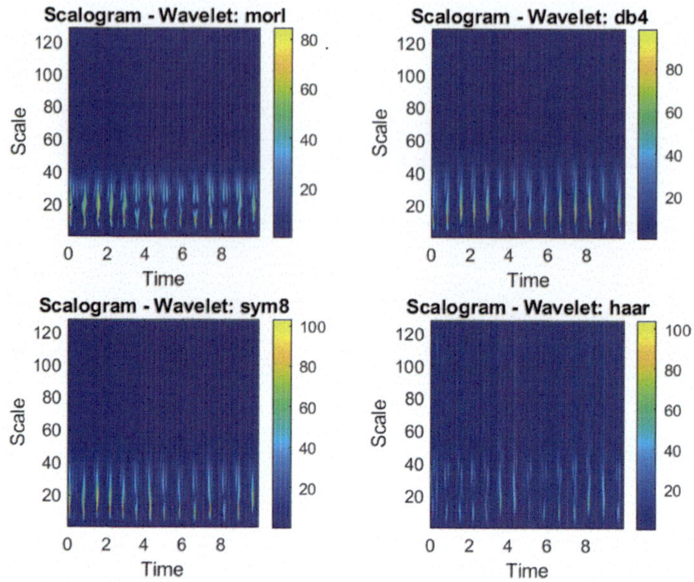

Fig. 9 Observed signal

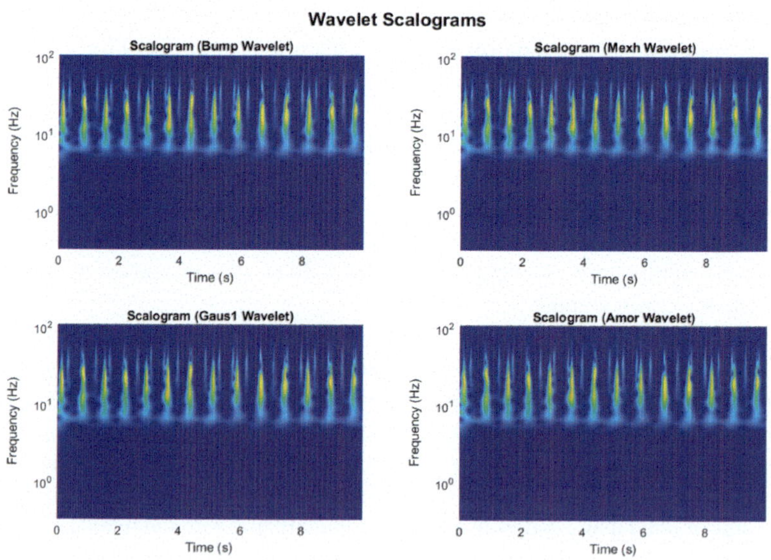

Fig. 10 Observed signal

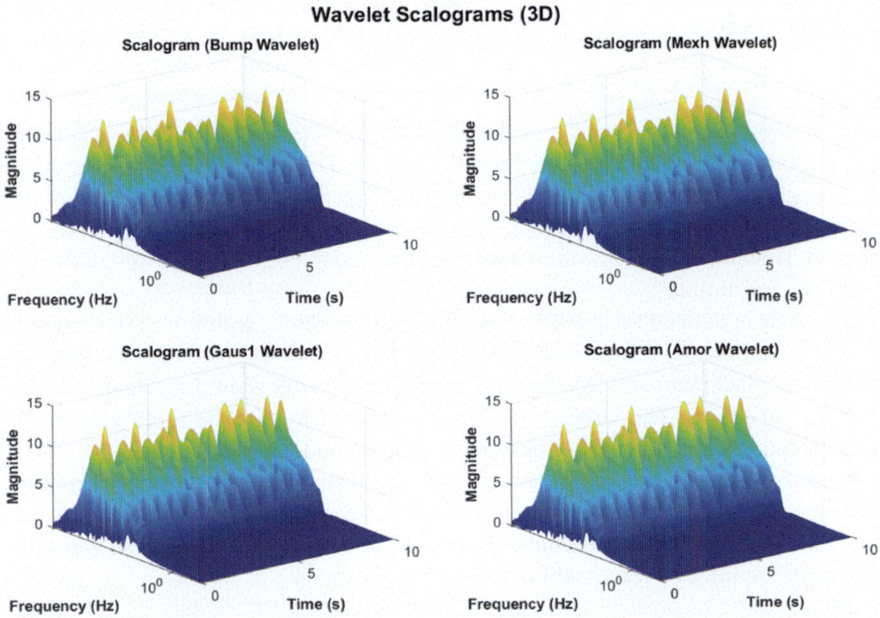

Fig. 11 Observed signal

information from the mixed fetal and maternal ECG signal. The Symlet wavelet demonstrates excellent time and frequency localization properties, making it highly suitable for capturing the distinct features and complex patterns present in the signal.

Furthermore, by employing a scale interval ranging from 1 to 64, the Symlet wavelet offers a wide range of scales that effectively captures both the fine details of fetal activity and the broader patterns of maternal activity. This enables a comprehensive analysis of the signal, resulting in enhanced detection and characterization of fetal and maternal components [8–11].

These findings highlight the significance of selecting an appropriate wavelet for analyzing complex physiological signals like mixed fetal and maternal ECG. The Symlet wavelet, with its remarkable localization properties and wide range of scales, emerges as a valuable choice for extracting relevant information and enabling optimal analysis of such signals, Furthermore, wavelets play a vital role in energy efficiency and electrical machines by enabling precise signal analysis and optimizing performance. Their ability to capture localized frequency information allows for efficient fault detection, condition monitoring, and improved energy consumption patterns [12–14].

4 Conclusion

In conclusion, the comparative analysis of spectrograms and scalograms for ECG signal analysis has revealed the trade-offs between time and frequency resolution. Spectrograms, generated using the Short-Time Fourier Transform (STFT), provide a time-frequency representation of the signal but suffer from poor frequency resolution and lack of fine detail. On the other hand, scalograms, computed using Continuous Wavelet Transform (CWT), offer improved time-frequency localization and better frequency resolution.

The study examined various window types and lengths, as well as hop sizes, within the STFT framework. It was observed that different window types, such as Hanning, Hamming, Blackman, and Rectangular, have distinct effects on the visualization of ECG spectrograms. The choice of window type can impact the representation of QRS complexes and other features related to fetal and maternal activities.

Additionally, the investigation involved exploring different wavelet families, including Morlet, Daubechies, Symlet, Haar, Coiflet, Meyer, Mexican Hat, Shannon, BiorSplines, Gabor, and amor. The results demonstrated that certain wavelets, such as the Symlet wavelet, exhibit superior performance in capturing the intricate details and patterns present in mixed fetal and maternal ECG signals.

Overall, the study highlights the importance of considering the trade-offs between time and frequency resolution in ECG signal analysis. While spectrograms provide a global overview of time-frequency content, scalograms offer better localization and resolution for analyzing specific features. The choice of window type, wavelet family, and parameters like window length and hop size should be carefully considered based on the specific requirements of the analysis.

One original contribution of this study is the comprehensive comparative analysis of spectrograms and scalograms for ECG signal analysis. While previous studies have explored these techniques individually, this research systematically investigates the trade-offs between time and frequency resolution in the context of ECG analysis. By examining various window types, wavelet families, and parameter settings, the study provides valuable insights into the strengths and limitations of spectrograms and scalograms in capturing important features of mixed fetal and maternal ECG signals. Future research could focus on exploring advanced techniques that combine the strengths of both spectrograms and scalograms, aiming to achieve a balance between time and frequency resolution. This would involve developing hybrid methods or adaptively selecting window types and wavelets based on the characteristics of the ECG signal. Such advancements would further enhance the analysis and interpretation of ECG signals, leading to improved detection and understanding of fetal and maternal activities.

References

1. Ziani, S., El Hassouani, Y.: A new approach for extracting and characterizing fetal electrocardiogram. Traitement du Signal, vol. 37, no. 3, pp. 379–386 (2020). https://doi.org/10.18280/ts.370304
2. Jamshidian-Tehrani, F., Sameni, R., Jutten, C.: Temporally nonstationary component analysis; application to noninvasive fetal electrocardiogram extraction. IEEE Trans. Biomed. Eng. **67**(5), 1377–1386 (May 2020). https://doi.org/10.1109/TBME.2019.2936943
3. Ziani, S., El Hassouani, Y., Farhaoui, Y.: An NMF based method for detecting RR interval. In: International Conference on Big Data and Smart Digital Environment. Springer (2019)
4. De Moor, B., De Gersem, P., De Schutter, B., Favoreel, W.: Sep DAISY: a database for fiction of systems. J. A, Special Issue on CACSD (Comput. Aided Control Syst. Des.), **38**(3), 4–5 (1997)
5. Tang, S.-N.: Design of a STFT hardware kernel for spectral Doppler processing in portable ulrasound systems. In: 2016 IEEE 5th Global Conference on Consumer Electronics, Kyoto, Japan, 2016, pp. 1–2 (2016). https://doi.org/10.1109/GCCE.2016.7800466
6. Aakaaram, V., Bachu, S.: MRI and CT image fusion using synchronized anisotropic diffusion equation with DT-CWT decomposition: smart technologies, communication and robotics (STCR). Sathyamangalam, India **2022**, 1–5 (2022). https://doi.org/10.1109/STCR55312.2022.10009173
7. Qassim, Y.T., Cutmore, T., Rowlands, D.: Multiplier Truncation in FPGA Based CWT. In: International Symposium on Communications and Information Technologies (ISCIT). Gold Coast, QLD, Australia, vol. 2012, pp. 947–951 (2012). https://doi.org/10.1109/ISCIT.2012.6381041
8. Wang, M., et al.: Low-latency in situ image analytics with FPGA-based quantized convolutional neural network. IEEE Trans. Neural Netw. Learn. Syst. **33**(7), 2853–2866 (2022). https://doi.org/10.1109/TNNLS.2020.3046452
9. Ziani, S.: Contribution to single-channel fetal electrocardiogram identification. Traitement du Signal, vol. 39, no. 6, pp. 2055–2060 (2022). https://doi.org/10.18280/ts.390617
10. Ziani, S., Farhaoui, Y., Moutaib, M.: Extraction of fetal electrocardiogram by combining deep learning and SVD-ICA-NMF methods. Big Data Min. Anal. **6**(3), 301–310 (2023). https://doi.org/10.26599/BDMA.2022.9020035
11. Ziani, S.: Fetal electrocardiogram identification using statistical analysis. In: Farhaoui, Y., Rocha, A., Brahmia, Z., Bhushab, B. (eds.) Artificial Intelligence and Smart Environment. ICAISE 2022. Lecture Notes in Networks and Systems, vol. 635. Springer, Cham (2023). https://doi.org/10.1007/978-3-031-26254-8-64
12. Chaou, Y. et al.: Nonlinear control of the permanent magnet synchronous motor PMSM using backstepping method. WSEAS Trans. Syst. Control **17**, 56–61 (2022). https://doi.org/10.37394/23203.2022.17.7
13. Ouhadou, M. et al.: Experimental modeling of the thermal resistance of the heat sink dedicated to SMD LEDs passive cooling. In: Proceedings of the 3rd International Conference on Smart City Applications (2018)
14. Ben, Hafid, et al.: Permanent magnet synchronous motor PMSM control by combining vector and PI controller. WSEAS Trans. Syst. Control **17**, 244–249 (2022)

Assessing the Environmental Effects Associated with Flood Protection Measures

Najlaa Ben-Lhachemi, Mohammed Benchrifa, Souad Nasrdine, Jamal Mabrouki, and Miloudia Slaoui

Abstract Environmental deterioration is brought on by both short-term and long-term variations in the components of the air, water, and soil levels. Air pollution is significantly impacted by construction operations. The purpose of this study is to evaluate the effects of flood protection efforts made at Oued Ikkour in the Zaouiat Cheikh area. We used the AirQ+ software, focusing on the PM10 and O_3 pollutants, to estimate the health-related impacts of air pollution. To display the distribution of pollutants and do statistical analysis on PM10 and O_3 concentrations, we again used the SPSS program. Our research sought to thoroughly assess the effects of air pollution brought on by the flood protection activities at Oued Ikkour. We thoroughly examined the effect on public health and presented a thorough statistical study of PM10 and O_3 pollutant levels by using the appropriate tools and methodology.

Keywords Air pollution · Environment · Particulate matter PM10 · Ozone O_3 · AirQ+ · Pollutants · SPSS

N. Ben-Lhachemi (✉) · S. Nasrdine · M. Slaoui
Energy, Materials and Sustainable Development (EMDD) Laboratory—Higher School of Technology—SALE, Center of Water, Natural Resources, Environment and Sustainable Development (CERN2D), University Mohammed V in RABAT, 10100 Rabat, Morocco
e-mail: benlhacheminajlaa@gmail.com

M. Benchrifa
Solar Energy and Environment Team, Faculty of Science, Mohammed V University in Rabat, Rabat, Morocco

J. Mabrouki
Laboratory of Spectroscopy, Molecular Modeling, Materials, Nanomaterials, Water and Environment, Faculty of Science, CERNE2D, Mohammed V University in Rabat, Rabat, Morocco

J. Mabrouki and M. Azrour (eds.), *Integrated Solutions for Smart and Sustainable Environmental Conservation*, Studies in Systems, Decision and Control 527,
https://doi.org/10.1007/978-3-031-55787-3_9

1 Introduction

Faced with the global environmental crisis that threatens our planet, it is imperative to study in depth various human activities such as the combustion of fossil fuels, industrial discharges and construction works against flooding and their impact on the atmospheric environment [1].

Flood protection works consist of protecting vulnerable areas against overflowing watercourses [2, 3]. Concrete is often used the most in the construction of dams, levees and other flood protection infrastructure [4]. The production of cement, the main component of concrete, is a major source of carbon dioxide (CO_2) emissions into the atmosphere, which contributes to air pollution [5]. In addition, construction work often requires the use of vehicles and heavy equipment such as trucks, forklifts, etc. These vehicles can raise fine dust containing PM2.5 and PM10 particles when traveling on unpaved surfaces or when carrying out loading and unloading operations. These pollutants can have adverse effects on air quality and human health as well as on surrounding ecosystems [6, 7].

Flood prevention attempts to save metropolitan areas from floods, but by shifting flood risks to weaker populations, it frequently contributes to environmental injustice [8, 9]. This can cause displacement and has a disproportionately negative impact on underprivileged populations. An illustration of this injustice is the Taipei Area Flood Control Plan. In order to address this, a paradigm change toward flood adaptation is suggested as a more just method of reducing flood risks [10].

The implementation of flood protection measures in stream catchments and protected areas is essential to safeguarding settlements and economically significant locations against flooding [11, 12]. Focusing on protection in the immediate region or close by is important if these precautions are not practical. Environmental impact studies assist in choosing the optimum flood protection option while taking into account the environmental risk [13].

There are various methods available for flood protection [14]. One approach is putting in place flood protection systems in stream catchments and protected areas, which reduce peak flow and spread floodwaves over a longer period of time. Focusing on protection in the immediate area or close by can be required if this is not feasible or effective. When there is little time between the detection of a flood threat and its commencement, expanding the watercourse's capacity can be used in conjunction with other flood mitigation techniques [15, 16]. The choice of the best flood prevention strategy ultimately depends on variables like hydraulic modeling and environmental effect analysis [17]. In order to advance conventional approaches, Yang et al. introduce a new flood control model with operational principles. For effective solutions, it makes use of two-stage linear programming (TSLP) and mixed integer linear programming (MILP). Results demonstrate that while TSLP is over 150 times faster, MILP decreases rule violation by an average of 17%. The three Gorges reservoir, which holds 76% of floodwater during once-in-a-century floods, contributes significantly to flood detention [18].

Cost–benefit analyses frequently make it impractical to achieve full protection just through structural flood defense techniques [19]. Such actions must be balanced against their potential advantages and the costs of their implementation and upkeep [20]. Consequently, a balanced strategy that includes structural and non-structural solutions [21], taking into account cost-effectiveness and long-term sustainability, is often more useful and efficient in managing flooding [22]. In order to supplement the physical structures and advance comprehensive flood risk reduction, non-structural approaches concentrate on enhancing resilience and reducing vulnerability [23, 24].

This chapter will specifically analyze the impact of Oued Ikkour flood protection works on air quality, soil and population health. For this, we will use the AirQ+ software to assess the effects on population health, as well as the SPSS software to conduct a statistical study of PM10 and O_3 concentration data. This approach will allow us to provide a complete and rigorous assessment of the environmental consequences of this work and its impact on health.

2 Methodology

2.1 Description of the Study Area: Oued Ikkor Zaouiat Sheikh

The city of Zaouiat Cheikh is located at the foothills of the Middle Atlas Mountains in Morocco, 69 km from the city of Béni Mellal. Its latitude is approximately 32.6541, and its longitude is approximately −5.92138. This region is known for its Mediterranean climate with hot summers.

Zaouiat Cheikh has been the subject of a flood protection study, as part of a comprehensive study of 11 sites distributed between the provinces of Fkih Ben Salah and Béni Mellal. Oued Ikkour, which drains a watershed of approximately 49 km^2, flows through the center of the city for a distance of about 1.5 km, passing under the RN 8 through a reinforced concrete culvert measuring 4.00 m × 4.00 m. No signs of degradation were observed on this structure.

The proposed development scheme for Oued Ikkour includes the following measures: construction of a protective wall with cut stones (from Pk 0 + 190 to Pk 0 + 380), installation of gabion structures on the left bank between PK 0 + 547 and PK 0 + 664, strengthening of the existing hydraulic structure under RN8 with another opening measuring 5 m × 4 m, construction of a trapezoidal concrete canal between the two hydraulic structures (between Pk 0 + 877 and Pk 1 + 158), and construction of a wall with cut stones (from Pk 1 + 503 to Pk 1 + 649).

Regarding Oued Ikkor, the development plan involves the canalization of the watercourse and the construction of two weirs upstream and downstream. The canalization will be carried out in two trapezoidal sections with an earth foundation and concrete slopes. The first section is located upstream of the RN8 crossing, and the second section is downstream (Table 1).

Table 1 Details of trapezoidal canalization sections with earth foundation and concrete slopes

	Upstream section	Downstream section
Length	811 m	733 m
Section	Trapezoidal	Trapezoidal
Width of the foundation	13 m	11 m
Height of the lining	2.3 m	2.5 m
Slope of the slopes	3H/2 V	3H/2 V
Longitudinal slope	0.5%	0.7%

The construction of several gabion drops is necessary to reduce the slopes of the river and the flow velocities within the canalized zone. In the upstream section, there are a total of 33 drops, one with a height of 0.5 m and 32 with a height of 1 m. In the downstream section, there are 19 drops, two with a height of 0.5 m and 17 with a height of 1 m.

We have worked with AirQ+, which is developed by the World Health Organization (WHO) aimed at assessing outdoor air quality and quantifying the impact of air pollution on health using this software is to conduct a study to assess the amount of contaminants related to respiratory diseases resulting from the dispersion of PM10 particles in the Zaouiat Cheikh area.

2.2 *AirQ+ Program*

AirQ+ was created by the WHO Regional Office for Europe to estimate the burden and effects of air pollution on human health [25] (Fig. 1).

To evaluate the immediate and long-term consequences of ambient air pollution, it offers simple approaches. Black carbon (BC), PM2.5, PM10, NO_2, and O_3 are among the contaminants that AirQ+ deals with. It may also calculate the impacts of indoor air pollution brought on by using solid fuels (SFU). The computations might take into account a variety of health markers related to both acute and chronic mortality and morbidity.

2.3 *IBM SPSS Platform*

IBM SPSS Statistics is an influential software platform for statistical analysis (Fig. 2).

It offers an intuitive user interface and a wide range of functions that let your company quickly extract insightful knowledge from your data. A well-informed decision can be made thanks to its sophisticated statistical techniques, which guarantee accurate findings. Data preparation, management, analysis, and reporting are all included in the software's comprehensive coverage of the analytics process.

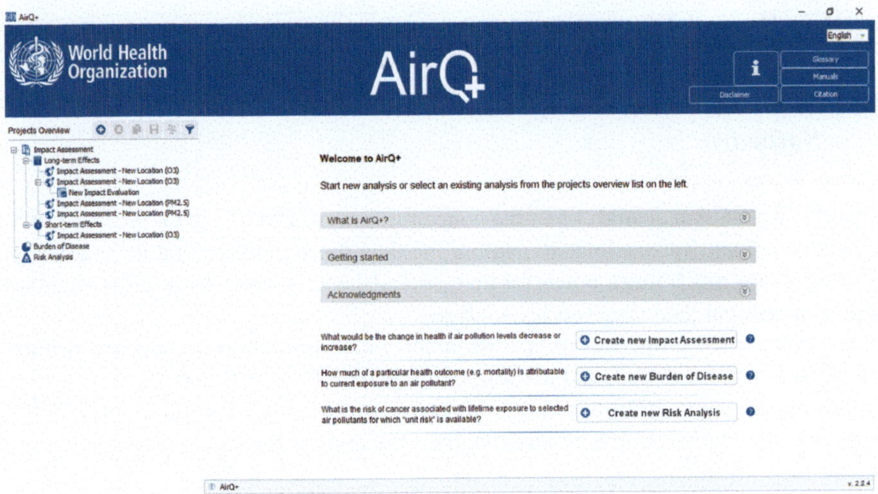

Fig. 1 Interface of AirQ+ program

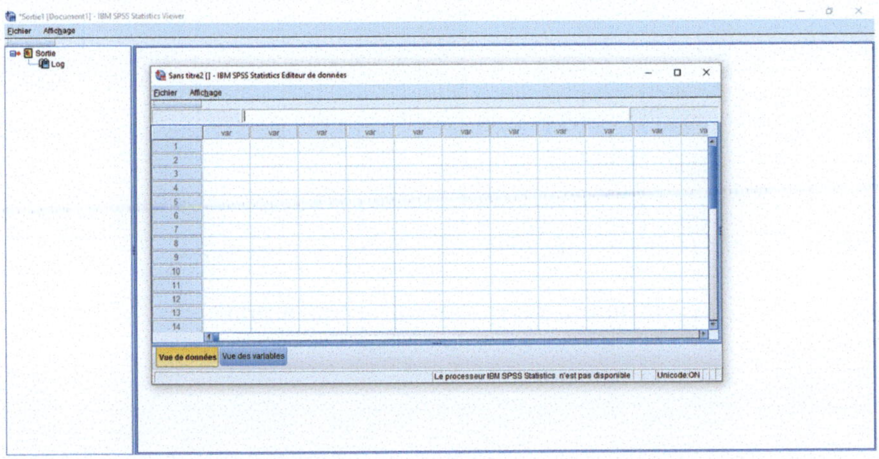

Fig. 2 Interface of SPSS statistical platform

3 Results and Discussion

3.1 Statistical Analysis of PM10 Concentrations with SPSS Software

To perform a statistical analysis of the concentrations of PM10 and O_3 we use SPSS, is a statistical software, allows to explore the data more in depth and more quickly, offering a tool much more powerful than spreadsheets, databases or tools standard multidimensional data reserved for analysts.

We defined the properties of two variables which are the average concentration and the months, then we found the following results (Figs. 3, 4 and 5).

During the years 2020 and 2021 there are respectively maximums of 172.189 (lg/m^2) at the end of summer (August), then the concentration decreased in winter to 54 and 45 (lg/m^2).

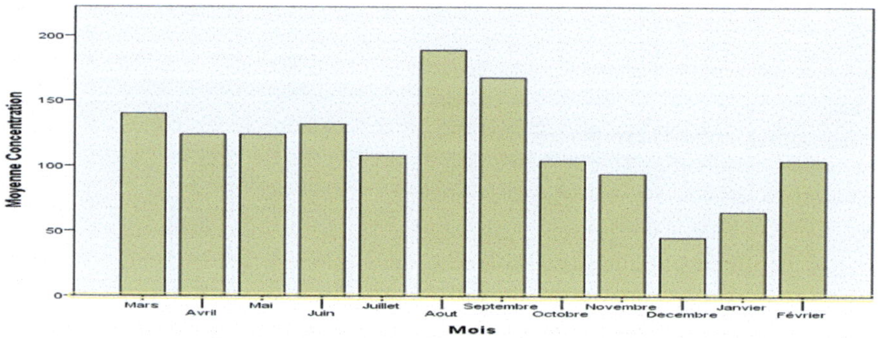

Fig. 3 Concentration of PM10 according to the months of the year 2020

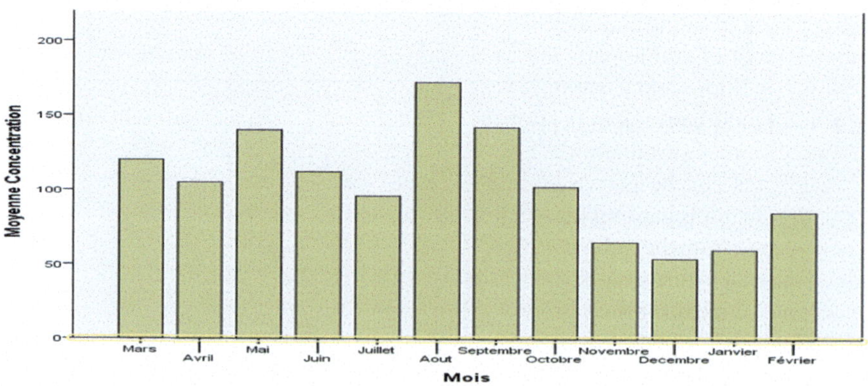

Fig. 4 Concentration of PM10 according to the months of the year 2021

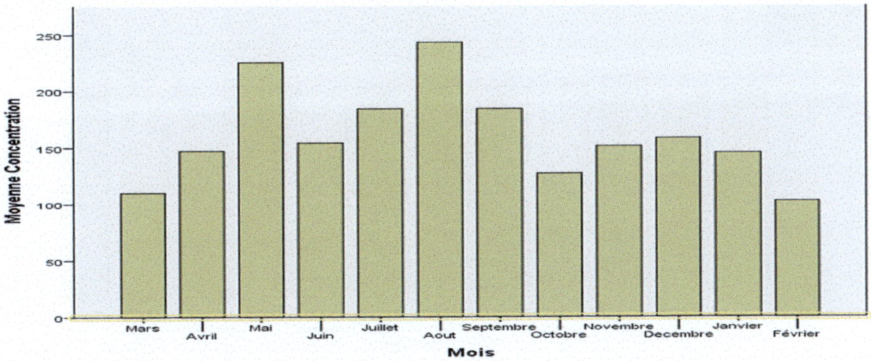

Fig. 5 Concentration of PM10 according to the months of the year 2022

The histogram represents the variation of O_3 concentrations according to the months during the year 2022 at the start of the flood protection works of Oued Ikkour Zaouiat Cheikh. We notice during the work, the maximum concentration of PM10 increases to 243 (lg/m^2), then decreases but this time gradually to 158 (lg/m^2). So we see that this work plays an important role in increasing the production of PM10 particles.

3.2 *Statistical Analysis of O_3 Concentrations with SPSS Software*

In the same way, we obtained the following histograms (Figs. 6, 7, and 8).

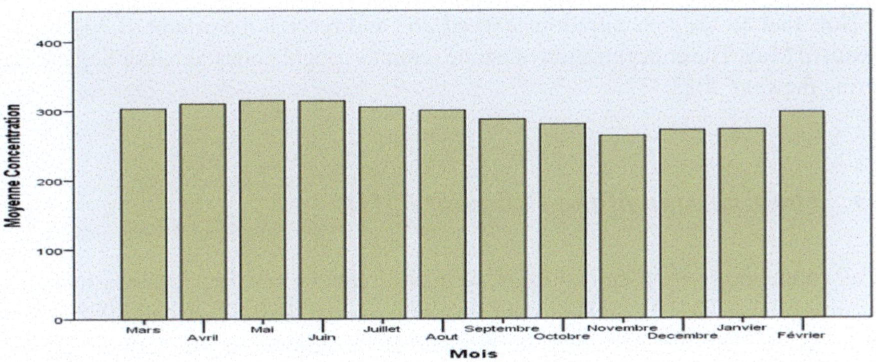

Fig. 6 Concentration of O_3 according to the months of the year 2020

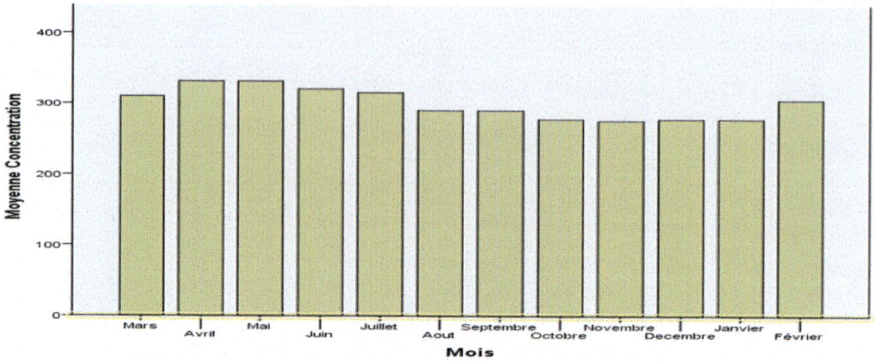

Fig. 7 Concentration of O_3 according to the months of the year 2021

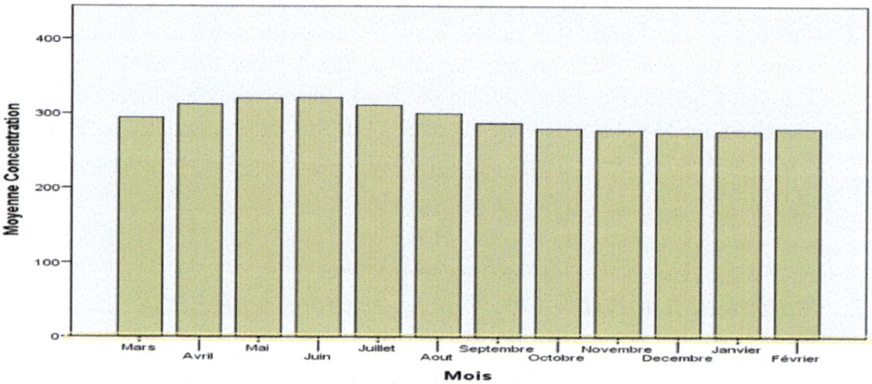

Fig. 8 Concentration of O_3 according to the months of the year 2022

Note that all O_3 concentrations exceed 263 and reach a maximum of 331 in the month of May. The concentration of ozone remains roughly constant after September during the year 2022.

3.3 *The Emission of the Pollutant PM10*

After conducting a thorough analysis, we have created a new impact assessment for the insertion of PM10 pollutant data for the 3 years; 2020, 2021 and 2022, After inserting the data, we specified the number of inhabitants of the Zaouiat Cheikh area and the determination of the cause (respiratory diseases), in order to obtain the following results.

Figure 9 represents the concentrations of the pollutant PM10 according to the cases of contamination. The concentration of PM10 gradually increases until it reaches a

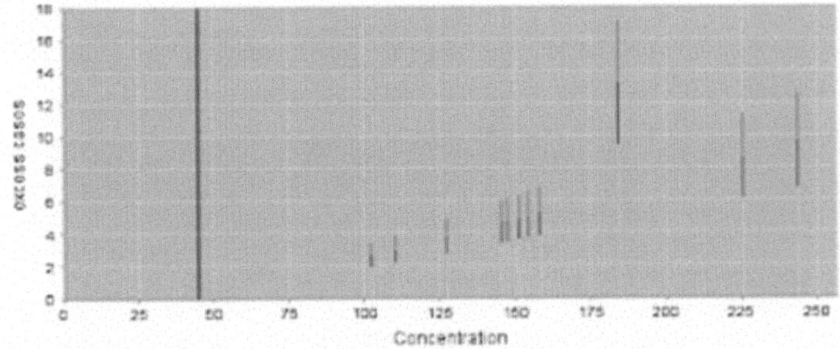

Fig. 9 Contamination as a function of PM10 concentration in 2020

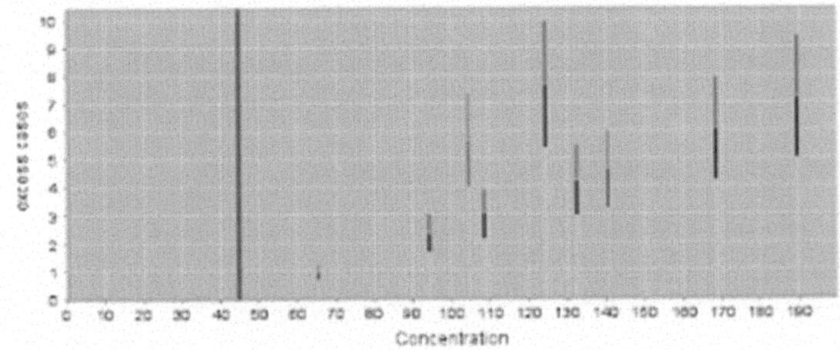

Fig. 10 Contamination as a function of PM10 concentration in 2021

maximum value of 172 for a number of cases which reaches a total of 8. This is due to the production of PM10 from human sources such as agricultural activities and residues of combustion.

Regarding Fig. 10, it is obvious that the maximum contamination of the populations reaches up to 10 cases because of the increase in human activities.

During the construction work against the flooding of Oued Ikkour, there is a sharp increase in the number of people who suffer from respiratory diseases due to the emission of the pollutant PM10. These emissions come from the use of construction machinery that generates dust containing PM10 and materials such as concrete (Fig. 11).

3.4 *The Emission of the Pollutant Ozone (O_3)*

In the same way we have inserted the ozone concentration data for the three years: 2020, 2021 and 2022.

Figures 12 and 13 show that the concentration of ozone increases gradually to a maximum of 315 $\mu g/m^3$ which can cause the contamination of 8 cases.

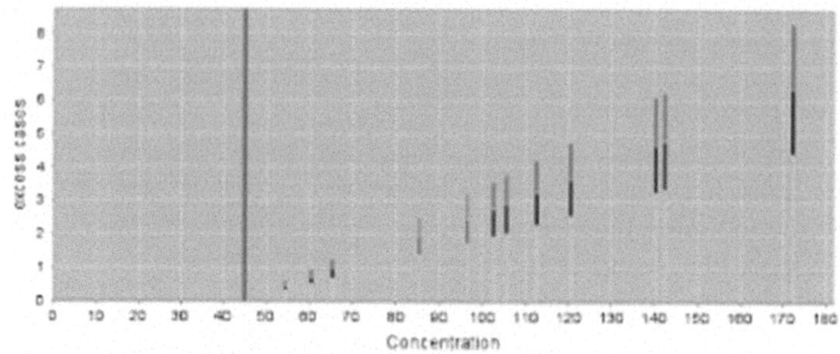

Fig. 11 Contamination as a function of PM10 concentration in 2022

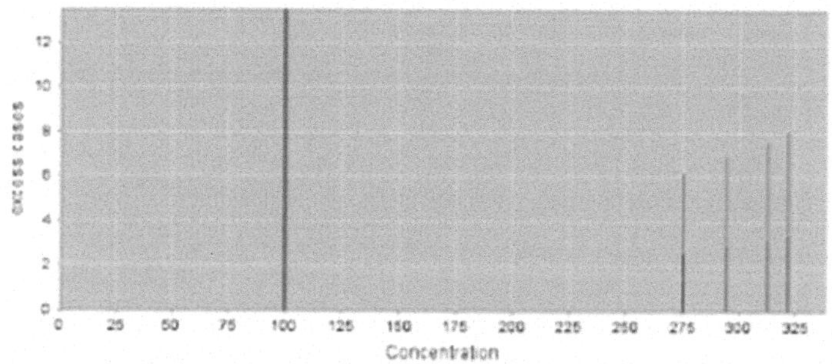

Fig. 12 Contamination as a function of O_3 concentration in 2020

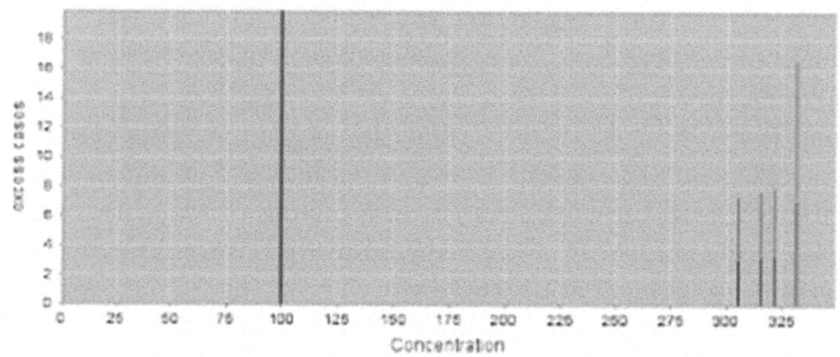

Fig. 13 Contamination as a function of O_3 concentration in 2021

The sudden increase in the concentration of ozone allows the increase in the number of contamination up to 17 cases.

The concentration of ozone is often higher during the summer months due to the chemical reaction between NOx and VOCs which come from several sources (transport, forest fire, industrial activities …) in the presence of sunlight this exposure to high levels of ozone contributes to the development of these respiratory diseases.

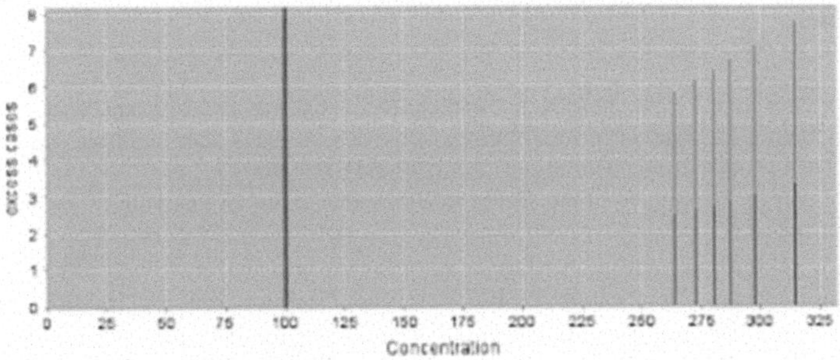

Fig. 14 Contamination as a function of O_3 concentration in 2022

The histogram above Fig. 14 shows the increase in the concentration of ozone due to the protection works against the flooding of Oued Ikkour with a total of 8 cases of contamination.

4 Conclusion

Our planet's air quality is a concern, and it is caused by both human activity and natural emissions from things like burning fossil fuels, manufacturing, and construction projects, as well as from things like volcanic eruptions, soil erosion, and raging forest fires.

Our research focused specifically on how flood protection measures affect air, soil, water, and human health. This prompted us to use the AirQ+ program to analyze the effects of the construction on human health and to conduct a statistical analysis with SPSS to determine the levels of PM10 and O_3 pollution produced before and after the construction.

It is time to take action to preserve our natural inheritance; environmental conservation cannot be seen as a choice. Therefore, by conducting a thorough environmental assessment to determine the consequences on the atmosphere, the impact of these activities on the environment can be significantly reduced through planning and design that respects the environment. By building ecological corridors and relocating native species that are impacted by the flood protection work, it is crucial to conserve the current fauna and flora.

References

1. Tin, T., Fleming, Z.L., Hughes, K.A., Ainley, D.G., Convey, P., Moreno, C.A., Pfeiffer, S., Scott, J., Snape, I.: Impacts of local human activities on the Antarctic environment. Antarct. Sci. **21**, 3–33 (2009). https://doi.org/10.1017/S0954102009001722
2. Maur, T., van de Giesen, N.: Floods and flood management in Pakistan. Phys. Chem. Earth Parts ABC **47–48**, 11–20 (2012). https://doi.org/10.1016/j.pce.2011.08.014
3. Owusu, S., Wright, G., Arthur, S.: Public attitudes towards flooding and property-level flood protection measures. Nat. Hazards **77**, 1963–1978 (2015). https://doi.org/10.1007/s11069-015-1686-x
4. Grigg, N.: Water Infrastructure and Equipment, pp. 227–240 (2016)
5. Lehne, J., Preston, F.: Making concrete change: innovation in low-carbon cement and concrete (2018)
6. Chen, S., Zhang, X., Lin, J., Huang, J., Zhao, D., Yuan, T., Huang, K., Luo, Y., Jia, Z., Zang, Z., Qiu, Y., Xie, L.: Fugitive road dust PM2.5 emissions and their potential health impacts. Environ. Sci. Technol. **53**, 8455–8465 (2019). https://doi.org/10.1021/acs.est.9b00666
7. Kalaiarasan, G., Balakrishnan, R.M., Sethunath, N.A., Manoharan, S.: Source apportionment studies on particulate matter (PM10 and PM2.5) in ambient air of urban Mangalore, India. J Environ. Manage. **217**, 815–824 (2018). https://doi.org/10.1016/j.jenvman.2018.04.040
8. Walker, G., Burningham, K.: Flood risk, vulnerability and environmental justice: evidence and evaluation of inequality in a UK context (2011). https://doi.org/10.1177/0261018310396149. Accessed 23 Jun 2023
9. Martynoga, C.: The Environmental Justice Implications of the Planning Policy and Practice of Flood Risk Management in New Zealand. The University of Waikato, Thesis (2018)
10. Liao, K.-H., Chan, J.K.H., Huang, Y.-L.: Environmental justice and flood prevention: The moral cost of floodwater redistribution. Landsc. Urban Plan. **189**, 36–45 (2019). https://doi.org/10.1016/j.landurbplan.2019.04.012
11. Jha, A.K., Bloch, R., Lamond, J.: Cities and Flooding: A Guide to Integrated Urban Flood Risk Management for the 21st Century. World Bank Publications (2012)
12. Heidenreich, A., Masson, T., Bamberg, S.: Let's talk about flood risk—evaluating a series of workshops on private flood protection. Int. J. Disaster Risk Reduct. **50**, 101880 (2020). https://doi.org/10.1016/j.ijdrr.2020.101880
13. Šoltész, A., Zeleňáková, M., Čubanová, L., Šugareková, M., Abd-Elhamid, H.: Environmental impact assessment and hydraulic modelling of different flood protection measures. Water **13**, 786 (2021). https://doi.org/10.3390/w13060786
14. Moel, H., Alphen, J.: Flood maps in Europe–methods, availability and use. Nat. Hazards Earth Syst. Sci. **9** (2009). https://doi.org/10.5194/nhess-9-289-2009
15. Ferguson, C.R., Fenner, R.A.: The potential for natural flood management to maintain free discharge at urban drainage outfalls. J. Flood Risk Manag. **13**, e12617 (2020). https://doi.org/10.1111/jfr3.12617
16. Dadson, S.J., Hall, J.W., Murgatroyd, A., Acreman, M., Bates, P., Beven, K., Heathwaite, L., Holden, J., Holman, I.P., Lane, S.N., O'Connell, E., Penning-Rowsell, E., Reynard, N., Sear, D., Thorne, C., Wilby, R.: A restatement of the natural science evidence concerning catchment-based 'natural' flood management in the UK. Proc. R. Soc. Math. Phys. Eng. Sci. **473**, 20160706 (2017). https://doi.org/10.1098/rspa.2016.0706
17. Doorn, N.: Rationality in flood risk management: the limitations of probabilistic risk assessment in the design and selection of flood protection strategies. J. Flood Risk Manag. (2014). Wiley Online Library. https://doi.org/10.1111/jfr3.12044. Accessed 23 Jun 2023
18. Yang, Z., Zheng, H., Feng, S., Chen, C., Wang, J.: Optimal multireservoir operation for flood control under constrained operational rules. J. Flood Risk Manag. **15**, e12825 (2022). https://doi.org/10.1111/jfr3.12825

19. Ward, P.J., Jongman, B., Aerts, J.C.J.H., Bates, P.D., Botzen, W.J.W., Diaz Loaiza, A., Hallegatte, S., Kind, J.M., Kwadijk, J., Scussolini, P., Winsemius, H.C.: A global framework for future costs and benefits of river-flood protection in urban areas. Nat. Clim. Change **7**, 642–646 (2017). https://doi.org/10.1038/nclimate3350
20. Kind, J.M.: Economically efficient flood protection standards for the Netherlands (2014). https://doi.org/10.1111/jfr3.12026
21. Schanze, J., Penning-Rowsell, E., Nachtnebel, H., Meyer, V., Königer, P., Hutter, G., Olfert, A., Parker, D., Harries, T., Werritty, A., Neuhold, C., Holzmann, H.: Systematisation, evaluation and context conditions of structural and non-structural measures for flood risk reduction. FLOOD-ERA Joint Report (2014)
22. Shah, M.A.R., Rahman, A., Chowdhury, S.H.: Challenges for achieving sustainable flood risk management. J. Flood Risk Manag. **11**, S352–S358 (2018). https://doi.org/10.1111/jfr3.12211
23. Almoradie, A., de Brito, M.M., Evers, M., Bossa, A., Lumor, M., Norman, C., Yacouba, Y., Hounkpe, J.: Current flood risk management practices in Ghana: Gaps and opportunities for improving resilience. J. Flood Risk Manag. **13**, e12664 (2020). https://doi.org/10.1111/jfr3.12664
24. Ingwe, R.: Flood resilience and sustainable development in urban Nigeria: Integrating traditional and non-structural methods of mitigating and adapting to flooding in Cross River State, Southeastern Nigeria. Riskuri Și Catastr Nr **11**, 127–143 (2012)
25. WHO tool for behavioural insights on COVID-19. https://www.who.int/europe/tools-and-toolkits/airq-software-tool-for-health-risk-assessment-of-air-pollution. Accessed 15 Jun 2023

Investigation on the Phytosociology and Soil Characteristics of Carob Tree (*Ceratonia siliqua* L.) in Northeast Morocco: Implications for Biodiversity Conservation and Ecosystem Restoration

Yassine Mouniane, Ahmed Chriqui, Issam El-Khadir, Ayoub Bensaid, Rida Arabi, Mohamed Haitam Zoubir, Jamal Mabrouki, and Driss Hmouni

Abstract Biodiversity conservation relies on a thorough understanding of species and ecosystems. Phytosociology and soil studies are essential tools for comprehending plant communities and soil properties, enabling the development of effective strategies for preservation and restoration of ecosystems. In this study, we focused on the Carob tree (*Ceratonia siliqua* L.), a promising species for reforestation projects in mountainous areas. Its ability to adapt to different soil types and its ecological significance make it an attractive choice for ecosystem restoration. In this study conducted in northeast Morocco, phytosociology revealed the presence of several plant species associated with the Carob tree, contributing to biodiversity and ecosystem balance. Soil analysis provided detailed information on physicochemical characteristics, including a slightly alkaline pH, high electrical conductivity, significant organic matter content, and considerable levels of NPK nutrients. Additionally, trace mineral analysis showed the beneficial presence of iron, magnesium, zinc, and nickel, along with a moderate amount of sodium, while the toxic element cadmium was absent from the soil. Moreover, we performed X-ray diffraction (XRD) analysis to further characterize the structural properties of the soil. The XRD analysis revealed the presence of calcite and quartz peaks. These results offer valuable insights into the species associated with the Carob tree and facilitate the development of effective strategies for biodiversity conservation and restoration of endangered ecosystems.

Y. Mouniane (✉) · A. Chriqui · I. El-Khadir · A. Bensaid · R. Arabi · M. H. Zoubir · D. Hmouni
Laboratory of Natural Resources and Sustainable Development, Faculty of Sciences, Ibn Tofaïl University, KENITRA-University Campus, 14000 Kenitra, Morocco
e-mail: yassine.mouniane@uit.ac.ma

J. Mabrouki
Laboratory of Spectroscopy, Molecular Modelling, Materials, Nanomaterials, Water and Environment, Faculty of Science, CERN2D, Mohammed V University in Rabat, AV Ibn Battouta, Agdal, 10106 Rabat, Morocco

J. Mabrouki and M. Azrour (eds.), *Integrated Solutions for Smart and Sustainable Environmental Conservation*, Studies in Systems, Decision and Control 527,
https://doi.org/10.1007/978-3-031-55787-3_10

Keywords Phytosociology · Soil · North-east Morocco · Carob tree · *Ceratonia siliqua* L.

1 Introduction

The growing recognition of the importance of ecosystem restoration as a cost-effective and impactful solution to address biodiversity loss and climate change is closely tied to the increasing significance of forests in the face of ongoing climate change [1–3]. Furthermore, within the realm of ecosystem restoration, the Carob tree (*Ceratonia siliqua* L.) stands out as a promising species for reforestation efforts, given its adaptability to various soil types and its ecological significance.

The carob tree (*C. siliqua*), commonly called kharroub, slaghoua and tikida in Moroccan dialect [4], a sclerophyllous leguminous belonging to the Caesalpinaceae sub-family, is a relevant component of the Mediterranean vegetation. It has been cultivated for restoration of degraded arid areas and for human and animal consumption [5–26]. The cultivation of the carob tree holds both environmental and economic value [11] Furthermore, the species has received significant attention in recent times and has become economically important, with its pods and seeds being utilized as raw materials in the food, pharmaceutical, and cosmetics industries [9–28].

Currently, wild and naturalized carob trees are mainly found in the same geographical and climatic zone as the varieties cultivated in the Mediterranean region [11]. A phytosociological study of Mediterranean carob habitats highlighted their high floristic diversity, notably in the Western basin where the carob tree meets its maximum ecological gradient [12]. In addition to its ecological importance, carob habitat conservation could be of great importance for the future since carob tree is known for its tolerance to marginal soils drought conditions [15–22]. However, despite the longstanding interest of botanists and the significant cultural importance of the carob tree in Mediterranean human culture, there is still a lack of comprehensive surveys on the plant diversity associated with this tree [11–23].

In this study, we focused on the plant communities associated with the carob tree (*C. siliqua* L.) in Mount Moulay Abdeslam Rebbeta located in the north-western region of Morocco. In parallel with the phytosociological study, we focused on soil characterization in areas where carob is present. This involved analyzing the physico-chemical properties of the soil, as well as carrying out mineral analysis and X-ray diffraction (XRD) to gain an overall understanding of soil composition.

2 Material and Methods

2.1 Study Area

This study was conducted in the natural region of Mount Moulay Abdeslam Rebbeta, which belongs to the territorial commune Brikcha located in the Ouazzane region of northwest Morocco (Fig. 1). The region benefits from a sub-humid Mediterranean climate with a dry summer season with temperatures ranging between 19 and 32 °C and a cold, wet winter with temperatures ranging between 6 and 14 °C. The study area is renowned for its suitability for the growth of the carob tree, characterized by uneven relief, hills, and low mountains (Fig. 2a).

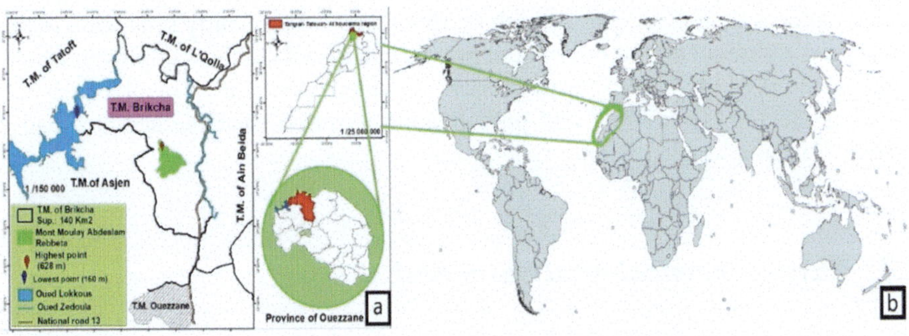

Fig. 1 **a** Location map of the study area; **b** positioning Morocco in the world

Fig. 2 **a** Carob tree (*C. siliqua* L.)—Photo of the Carob tree in its natural habitat; **b** pods of carob tree; **c** seeds of carob

2.2 Vegetation Analysis

The phytosociological surveys were carried out using a systematic sampling method, ensuring coverage of the different habitats present on Mont Moulay Abdeslam. Plant cover was assessed during the phytosociological surveys using the Braun-Blanquet method. This widely recognized method enabled us to visually estimate the abundance of plant cover in each survey [13]. The plot sizes ranged between 70 and 100 m^2 and the altitudinal range was between 500 and 628 m.

2.3 Soil Sampling

Soil samples were collected to a depth of 15 cm and subsequently processed in the laboratory. Subsequently, a pre-treatment was conducted on the samples in accordance with ISO 11464:2006 for physicochemical analyses, which involved drying, grinding, sieving, separation, and pulverization, ensuring that the samples had a particle size smaller than 2 mm.

2.4 Physicochemical Analysis of Soil

The physicochemical analyses of the soil were conducted using standard methods. The soil pH was measured in a diluted 1:5 (V/V) suspension using a pH meter, according to the ISO 10390:1994 [16]. The electrical conductivity of the soil is measured using a conductivity meter [29]. The soil organic matter content was determined following the methodology described by Enang [17].

2.5 Minerals Content

The quantification of nitrogen was performed using the Kjeldahl method [14]. The available phosphorus content was determined using the Olsen method [21]. The concentrations of total trace metal elements (Cd, Fe, K, Mg, Na, Ni, and Zn) were analyzed using Coupled Plasma Mass Spectrometry (ICP-MS) (X Series, Thermo Electron Corporation).

2.6 Structural Characterization by XRD

X-ray diffraction (XRD) analysis of the samples was performed using an X'Pert3 Powder diffractometer, using Kα radiation from copper (Kα = 1.5406 Å). Diffractograms were acquired in the 10–70° range in 2θ with a step size of 0.02° and an acquisition time of 0.5 s.

3 Results

3.1 Phytosociological

Our results indicate that there is a high diversity of plant species associated with carob in our study. Interestingly, these species belong to different botanical families, which shows the richness of the plant community associated with carob tree. This table provides a clear overview of the plant species associated with carob in the study, with specific families and species (Table 1).

Our findings suggest that a wide variety of plant species are related to carob in our research. It's interesting to note that these species are members of various botanical families, demonstrating the diversity of the carob tree-associated plant population. These latter have a significant role in the overall operation of the soil through its physical, biological, and chemical components, which has significant ramifications for soil fertility [20]. In our study area, the most common species of tree associated

Table 1 Species to be associated with the carob tree

Species name	Family	Habit	Life form	Bioclimatic stage	Photo
Viburnum tinus	Adoxaceae	Shrub	Perennial	SA, SH, H	Figure 3a
Pistacia lentiscus	Anacardiaceae	Shrub	Perennial	SA, SH, H	Figure 3b
Cistus albidus	Cistaceae	Shrub	Perennial	SA, SH, H	Figure 3c
Cistus salviifolius	Cistaceae	Shrub	Perennial	SA, SH, H	Figure 3d
Cistus monspeliensis	Cistaceae	Shrub	Perennial	SA, SH, H	Figure 3e
Arbutus unedo	Ericaceae	Tree	Perennial	SA, SH, H	Figure 3f
Calicotoma villosa	Fabaceae	Shrub	Perennial	SA, SH	Figure 3g
Quercus coccifera	Fagaceae	Tree	Perennial	SA, SH, H	Figure 3h
Quercus rotundifolia	Fagaceae	Tree	Perennial	SA, SH, H	Figure 3i
Rhamnus alaternus	Rhamnaceae	Shrub	Perennial	SA, SH, H	Figure 3j
Phillyrea latifolia	Oleaceae	Shrub	Perennial	SA, SH, H	Figure 3k
Olea europaea subsp. Oleaster	Oleaceae	Tree	Perennial	SA, SH, H	Figure 3l

Abbreviation: *SA* Semi-arid; *H* Humid; *SH* Subhumid

Fig. 3 Photo of species associated with the carob tree on Mount Moulay Abdeslam Rebbeta

with caroubiers is the aok, and similar findings were observed in the agricultural parks of Azilal, Morocco [25], discovered that the most common species of tree associated with caroubiers is the Holm oak (*Quercus rotundifolia*). This association suggests a balanced interaction on carbon and nitrogen mineralization in the soil between carob and oak [6]. According to [12] we can see similarities in the common associations between carob and Mediterranean forest species such as *Pistacia lentiscus*, *Olea europaea*, *Smilax aspera*, *Quercus coccifera*, *Rhamnus* spp. and *Phillyrea* spp.

3.2 Physicochemical of Soil

The pH of the soil at the carob tree site is slightly alkaline (Table 2). This information is supported by the study conducted by [7], as well as the compliance with the Tunisian NT.106.03 standard. These standards define the acceptable pH range for agricultural and environmental purposes. The slightly alkaline pH indicates that the soil is not overly acidic and falls within the recommended range for optimal plant growth.

The electrical conductivity of the soil is a measure of the concentration of dissolved salts present. In the case of the soil associated with carob trees, the electrical conductivity value indicates a high concentration of dissolved salts. This finding aligns with the standard NT.106.03, which specifies that the electrical conductivity should be less than 7000 μs cm^{-1} for acceptable soil quality. The high electrical conductivity suggests that the soil may have elevated levels of salts, which can have implications for plant growth and nutrient availability.

Table 2 Physical and chemical characteristics of the studied soil

Soil parameter		Maximum limit (Tunisian standards)[a]
pH	7.55	6.5–8.5
Electrical conductivity (μs cm^{-1})	337	7 (dS/m)
Organic matter content (%)	7.83 ± 0.11	–

[a]Tunisian standard for re-utilization of wastewater in irrigation (NT.106.03)

The organic matter content is 7,83% means that the soil contains a relatively high amount of organic matter. Organic matter plays a vital role in soil fertility and overall soil health. It contributes to nutrient availability, water retention capacity, and soil structure. The presence of a higher organic matter content suggests that the soil may have better nutrient-holding capacity and improved moisture retention, which can be beneficial for plant growth and overall soil productivity.

3.3 *Minerals Content of Soil*

Nitrogen content is very low, indicating limited availability of this essential element (Table 3). Is an essential macronutrient deficient in most soils especially in arid and semi-arid regions [18]. The phosphorus levels in the soil are within a moderate range. Similarly, the presence of iron at moderate levels in the soil provides important functions necessary for the growth and development of the plant, but we have a low potassium concentration and magnesium. On the other hand, the high concentration of sodium in the soil can influence the availability of other nutrients. Nickel and zinc concentrations are within the optimum range, ensuring adequate levels for plant needs. Finally, cadmium concentration is below optimum values, which is positive as cadmium is a toxic element.

Table 3 Nutrients NPK and minerals content of soil

Parameter	Concentration	Interpretation
Nitrogen (N) g/kg	38.7 ± 2.15	<0.05 (Very low)[a]
Phosphorus (P) mg/kg	19.52 ± 1.02	Moderate (7.1–20.0)[b]
Potassium (K) mg/kg	535 ± 2.86	<5865 (low)[a]
Iron (Fe) mg/kg	3.50	Moderate (2.51–5.0)[b]
Magnesium (Mg) g/kg	1.61	<7.29 (low)[a]
Sodium (Na) mg/kg	940.1	> 6.9 (very high)[a]
Nickel (Ni) mg/kg	9.67	<50 (optimum values)[c]
Zinc (Zn) mg/kg	49.58	<300 (optimum values)[c]
Cadmium (Cd) mg/kg	0	<2 (optimum values)[c]

Source [a][24], [b][27], [c]maximum values: standard AFNOR U44-041 [8]

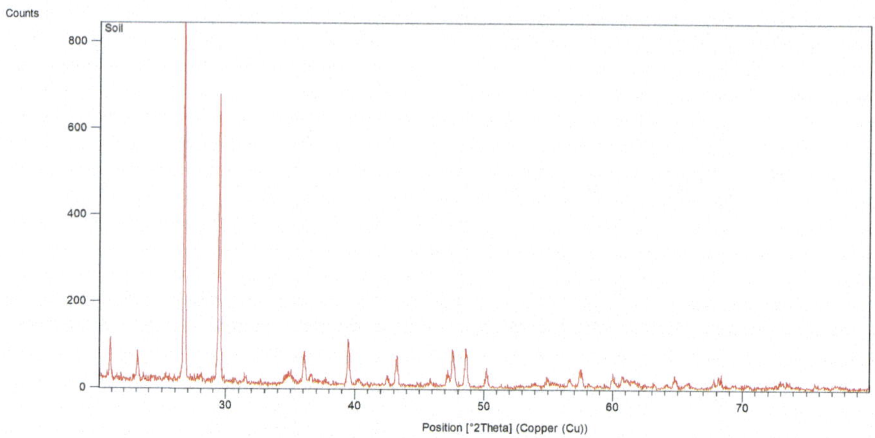

Fig. 4 Diffractogram of the studied soil

The results confirm carob's ability to thrive in relatively nutrient-poor soil conditions. Despite the limited presence of certain essential elements such as potassium, magnesium and nitrogen, the carob tree manages to thrive. This adaptability can be explained by the tree's mechanisms of stress tolerance and nutrient utilization efficiency.

3.4 *Structural Characterization by XRD*

The XRD analysis revealed the presence of two prominent peaks, corresponding to calcite and quartz (Fig. 4). The identification of calcite suggests that the soil may have originated from limestone rocks and potentially contains a significant amount of calcium, which is an essential nutrient for plant growth. The presence of quartz indicates a granitic or sandy composition of the soil, which can influence its physical properties such as permeability and structure.

4 Conclusion

In conclusion, this study is of great significance in highlighting the ability of carob to thrive in relatively nutrient-poor soil conditions. These results underline the potential of the carob tree for ecosystem restoration and biodiversity conservation, even in unfavorable environments. This discovery opens up new prospects for the use of carob in sustainable reforestation initiatives and testifies to its essential role in preserving our environment. It would therefore be interesting to continue studying these relationships to better understand ecological dynamics and opportunities for sustainable forest management.

References

1. Turner, W.R., Oppenheimer, M., Wilcove, D.S.: A force to fight global warming. Nature **462**(7271), 278–279 (2009)
2. Lewis, S.L., Wheeler, C.E., Mitchard, E.T.A., Koch, A.: Restoring natural forests is the best way to remove atmospheric carbon. Nature **568**(7750), 25–28 (2019)
3. Leclère, D., Obersteiner, M., Barrett, M., Butchart, S.H.M., Chaudhary, A., De Palma, A., DeClerck, F.A.J., Di Marco, M., Doelman, J.C., Dürauer, M.: Bending the curve of terrestrial biodiversity needs an integrated strategy. Nature **585**(7826), 551–556 (2020)
4. Aafi, A.: Le caroubier: Caractères botaniques et écologiques, groupements végétaux, techniques d'élevage en pépinière, traitements et soins culturaux, utilisation et production. Centre national de la recherche forestière, Maroc (1996)
5. Afif, M., Messaoud, C., Boulila, A., Chograni, H., Bejaoui, A., Rejeb, M.N., Boussaid, M.: Genetic structure of Tunisian natural carob tree (*Ceratonia siliqua* L.) populations inferred from RAPD markers. Ann. For. Sci. **65**(7), 710 (2008)
6. Aka, H., Darici, C.: Carbon and nitrogen mineralization in carob soils with Kermes oak and Aleppo pine leaf litter. Eur. J. Soil Biol. **41**(1–2), 31–38 (2005)
7. Baillie, I.C.: Soil survey staff 1999, soil taxonomy: a basic system of soil classification for making and interpreting soil surveys, agricultural handbook 436, Natural Resources Conservation Service, USDA, Washington DC, USA, pp. 869. Soil Use Manag. **17**, 57–60 (2001)
8. Baize, D.: Teneurs totales en métaux lourds dans les sols français premiers résultats du programme ASPITET. Unité de Science du sol. INRA, Ardon (1994)
9. Barracosa, P., Osorio, J., Cravador, A.: Evaluation of fruit and seed diversity and characterization of carob (*Ceratonia siliqua* L.) cultivars in Algarve region. Scientia Horticulturae **114**(4), 250–257 (2007)
10. Batista, M.T., Amaral, M.T., Proença Da Cunha, A.: Carob fruits as source of natural oxidant. In: Proceedings of the Communication in Third International Carob Symposium, 19–23. Tavira, Portugal (1996)
11. Batlle, I.: Carob tree: *Ceratonia siliqua* L.-Promoting the conservation and use of underutilized and neglected crops. 17. International Plant Genetic Resources Institute. Gatersleben, Germany (1997)
12. Baumel, A., Mirleau, P., Viruel, J., Bou Dagher Kharrat, M., La Malfa, S., Ouahmane, L., Diadema, K., Moakhar, M., Sanguin, H., Médail, F.: Assessment of plant species diversity associated with the carob tree (*Ceratonia siliqua*, Fabaceae) at the Mediterranean scale. Plant Ecol. Evol **151**, 185–193 (2018)
13. Braun-Blanquet, J.: Pflanzensoziologie: Grundzuege der Vegetationskunde; dritte, neubearbeitete und wesentlich vermehrte Auflage. Springer Verlag (1964)
14. Vasilas, B.L., Ham, G.E.: Nitrogen fixation in soybeans: an evaluation of measurement techniques. Agron. J. **76**(5), 759–764 (1984)
15. Correia, P.J., Martins-Loução, M.A.: The use of macronutrients and water in marginal Mediterranean areas: the case of carob-tree. Field Crops Res. **91**, 1–6 (2005)
16. Dawar, K., Fahad, S., Alam, S.S., Khan, S.A., Dawar, A., Younis, U., Danish, S., Datta, R., Dick, R.P.: Influence of variable biochar concentration on yield-scaled nitrous oxide emissions, Wheat yield and nitrogen use efficiency. Sci. Rep. **11**(1), 16774 (2021)
17. Enang, R.K., Yerima, B.P.K., Kome, G.K., Ranst, E.V.: Assessing the effectiveness of the Walkley-Black method for soil organic carbon determination in tephra soils of Cameroon. Commun. Soil Sci. Plant Anal. **49**(19), 2379–2386 (2018)
18. Herandez, L.E., Garate, A., Caroeba-Ruiz, R.: Effect of cadmium on the uptake, distribution and assimilation of nitrate in *Pisum sativum*. Plant Soil **189**, 97–106 (1997)
19. Hillcoat, D., Lewis, G., Verdcourt, B.: A new species of Ceratonia (Leguminosae-Caesalpinioideae) from Arabia and the Somali Republic. Kew Bull. **35**, 261–271 (1980)
20. Huber, G., Schaub, C.: La fertilité des sols: L'importance de la matière organique. Agriculture et terroir. Chambre d'agriculture Bas Rhin, France (2011)

21. Olsen, S.R.: Estimation of available phosphorus in soils by extraction with sodium bicarbonate. 939. US Department of Agriculture, United States (1954)
22. Ozturk, M., Dogan, Y., Sakcali, M.S., Doulis, A., Karam, F.: Ecophysiological responses of some maquis (*Ceratonia siliqua* L., *Olea oleaster* Hoffm. & Link, *Pistacia lentiscus* and *Quercus coccifera* L.) plant species to drought in the east Mediterranean ecosystem. J. Environ. Biol **31**, 233–245 (2010)
23. Ramón-Laca, L., Mabberley, D.J.: The ecological status of the carob-tree (*Ceratonia siliqua*, Leguminosae) in the Mediterranean. Bot. J. Linn. Soc. **144**, 431–436 (2004)
24. Shehu, B.M., Jibrin, M.J., Samndi, A.M.: Fertility status of selected soils in the Sudan Savanna Biome of Northern Nigeria. Int. J. Soil Sci. **10**(2), 74–83 (2015)
25. Taïbi, A.N., El Hannani, M., El Khalki, Y., Ballouche, A.: Les parcs agroforestiers d'Azilal (Maroc): une construction paysagère pluri-séculaire et toujours vivante. J. Alpine Res.| Revue de géographie alpine. **107**(3), (2019)
26. Tous, J., Romero, A., Batlle, I.: The carob tree: botany, horticulture, and genetic resources. Hortic. Rev. **41**, 385–454 (2013)
27. Ukabiala, M.E., Obazi, S.A.: Characterization and suitability evaluation of upland soils for Maize, Cassava and oil palm production in Kogi East, Nigeria. J. Soil Water Sci. **6**(1), 272–285 (2022)
28. Vourdoubas, J., Makris, P., Kefalas, J., Kaliakatsos, J.: Studies on the production of bioethanol from carob. In: The 12th National Conference and Technology Exhibition on Biomass for Energy. Industry and Climate Protection, Proceedings, pp. 489–493. Amsterdam (2002)
29. Walkley, A., Black, I.A.: An examination of the Degtjareff method for determining soil organic matter, and a proposed modification of the chromic acid titration method. Soil Sci. **37**(1), 29–38 (1934)

Phytoremediation of Contaminated Environments Using Halophytes: General Overview

Anas Tallou, Hassan Erraji, Yassine Al Masmoudi, Francesco Abbatantuono, Salem Alhajj Ali, Jamal Mabrouki, Khadija E. L. moustaqim, Francisco Pedrero Salcedo, and Gaetano Alessandro Vivaldi

Abstract Due to anthropogenic activities, potentially Toxic Elements (PTEs) have disrupted our ecosystems. These dangerous substances can infiltrate the food chain, affecting the environment and human life. To preserve our natural resources for future generations, scientists must look for eco-friendly alternatives. About 300 million tons of PTEs from industrial and consumer items, including Cr, Cu, Zn, As, Cd, Pb, and Sn, enter the receiving environments each year. Halophytes plants are found worldwide and grow in various places, including coastal zones, dunes, high-salinity environments, and inland deserts. They may also withstand PTEs stress due to evolved morphological and physiological characteristics such as reduced heavy metal intake through the root system. PTEs contamination causes agriculture yield losses, increases environmental concerns about human well-being, and causes a significant

A. Tallou (✉) · F. Abbatantuono · S. A. Ali · G. A. Vivaldi
Department of Soil, Plant and Food Sciences, University of Bari "Aldo Moro", Bari, Italy
e-mail: anas.tallou@uniba.it

H. Erraji
Laboratory of Bioresources, Biotechnology, Ethnopharmacology and Health, Faculty of Sciences, Mohamed Premier University, Oujda, Morocco

Renewable Energy and Energy Efficiency Training Institute, Oujda, Morocco

Y. Al Masmoudi
Laboratory of Geosciences and Environment Technics, Faculty of Sciences, Chouaib Doukkali University, 24000 El-Jadida, Morocco

J. Mabrouki
Laboratory of Spectroscopy, Molecular Modelling, Materials, Nanomaterials, Water and Environment, CERN2D, Mohammed V University in Rabat, Faculty of Science, AV IbnBattouta, Agdal, 10106 Rabat, Morocco

K. E. L. moustaqim
Improvement and Valuation of Plant Resources, Faculty of Sciences, Ibn Tofaïl University—KENITRA, 14000 Kenitra, Morocco

F. P. Salcedo
Center of Applied Biology and Soil Sciences (CEBAS), Murcia, Spain

J. Mabrouki and M. Azrour (eds.), *Integrated Solutions for Smart and Sustainable Environmental Conservation*, Studies in Systems, Decision and Control 527,
https://doi.org/10.1007/978-3-031-55787-3_11

drop in marine and coastal ecosystem biodiversity. Halophytes are plants that can flourish in extreme conditions such as high salinity and PTEs toxicity. This work emphasizes the role of the halophyte plants in terms of PTEs phytoremediation as a long-term strategy for polluted environmental ecosystem restoration, which aligns with the current European Green Deal strategy for a neutral Europe climate by 2050.

Keywords Phytoremediation · Halophytes · Waste management · Potentially toxic elements · Circular economy

1 Introduction

The Earth has approximately one billion hectares of saline land, and it is estimated that salt affects about 10% of the planet's total land area [63, 88]. It is also predicted that salt has affected the productivity of 10–50% of irrigated land (1.5 to 7 billion extra hectares). According to the FAO report (2017), the world's population is estimated to reach over 10 billion by 2050, increasing agricultural demand under a scenario of modest economic growth of around 50% over 2013 [63]. Furthermore, heavy metal pollution of the environment (recently called Potentially Toxic Elements (PTEs)) is a serious problem due to its harmful influence on these ecosystems. PTEs are a form of persistent pollutant that can be discovered in the environment. This class of pollutants has gained considerable attention because of its widespread dispersion, high hydrophobicity, long persistence, and negative effects on the environment and human health. Therefore, decreasing their influence is a major research priority [49, 65, 100].

PTEs have become a big concern worldwide [12]. They were discharged into the environment due to their extraction from ores and industries for diverse uses. The problem of PTEs contamination is growing more seriously as industrialization and disturbance of natural biogeochemical cycles increase. Unlike biological chemicals, PTEs are basically nonbiodegradable and accumulate in the environment. The existence of PTEs in soil and sediment found in streams poses a threat to both the environment and human well-being. These elements tend to build up in the bodily tissues of living organisms, a process known as bioaccumulation, and their levels rise as they move up the food chain, a phenomenon called biomagnification. PTEs present in soil can have harmful effects on soil microorganisms, potentially resulting in a decline in their population and activity [7].

The main sources of discharge of potentially toxic elements (PTEs) into the environment are industrial and mining activities, fossil fuel combustion, the improper disposal of industrial waste, and the excessive use of PTEs through fertilizers and pesticides. Traditional techniques for remediating PTEs encompass various approaches: (a) physical techniques like soil washing and extraction, as well as stabilizing PTEs in soil; (b) chemical techniques including vitrification technology, chemical leaching or fixation, and electrokinetic remediation; and (c) biological techniques such as phytoremediation or bioremediation utilizing bacteria, fungi, and plants [17]. Phytoremediation has been regarded as an inexpensive and more sustainable solution for removing or mitigating the effects of PTEs on the environment, and also it is likely to receive public support acceptance and the efficiency of the plants can be easily monitored [7, 27, 87]. It is widely recognized that halophytes contain more than one salinity tolerance mechanism, and this property has been employed in several agricultural research investigations [23]. Furthermore, it has been demonstrated that halophytes can resist large amounts of PTEs than glycophytes [6, 55]. Halophytes were examined extensively because of their resilience to survive in severe environments, with better outcomes for phytoremediation than salt-sensitive species [56]. They have the capacity to absorb and accumulate PTEs, resulting in more sustainable ecosystems with overall environmental properties [16, 74, 78]. Halophytes can be advantageous to agriculture in regions with saline soils or water. Farmers may cultivate halophytes in regions where other plants would die because of their properties.

This chapter introduces at first, the halophyte plants, their benefits and potential use and limitations. The second part will address the problem of PTEs pollution caused by different anthropogenic activities, and how these hazardous elements can enter the food chain and consequently threaten the ecosystem and human life. While the third part discusses the available remediation techniques to end up with the phytoremediation of polluted environments using halophytes.

2 Halophytes

Halophytes are important biological resources, accounting for around 1% of terrestrial angiosperms, with approximately 3000 species and 550 taxa dispersed across 120 families. Halophytes, as wild plants (e.g., Fig. 1), may adapt to various environmental conditions such as salt, drought, and extreme temperatures by employing operational procedures and processes dictated by their genetic code. Halophytes have the ability to uptake and transport significant quantities of salt to various parts of the plant. This is evident in the vibrant green and succulent aboveground biomass of many halophytic species, indicating their active functioning and adaptation to high salt loads, this begs the question of how these plants tolerate extreme salinity [31] (Christofilopoulos et al. 2016; Manousaki & Kalogerakis 2011a; Panta et al. 2014). Halophytes can be advantageous to agriculture in regions with saline soils or water. Farmers may cultivate halophytes in regions where other plants would die because of their properties.

Salinity-tolerant plants are classified as follows [31]:

(a) Obligatory halophytes (strictly halophytes, or halophytes): refer to plant species that exclusively thrive in saline environments. Some examples include *Salicornia herbacea, Salsola soda, Atriplex hastata, Plantago schwarzenbergiana,* and *Petrosimonia triandra.*

Fig. 1 Pink succulent Halophyte plant (*Salicornia europaea*)

(b) Facultative halophytes (adaptable halophytes, plants tolerant of salt): are species capable of growing in both saline and regular soils. Although they may exhibit limited growth in salty soils, they are more susceptible to desiccation during the dry season or periods of severe drought. Examples of facultative halophytes include *Lepidium ruderale, Poa bulbosa,* and *Matricaria chamomilla.*
(c) Halophobous: plants whose biomass declined as soil salinity increased.

Yensen and Biel [104] classified plants based on their potential to remediate saline soils as follows:

1. Excluders are plants that exclude salts at the root level. Given their limited ability to exclude salts, most terrestrial plants are excluders; they die if the salt load in their tissues increases. Some euhalophytes, such as the mangrove *Rhizophora spp.,* are salt-excluding plants, but their potential to remediate saline soil is restricted or non-existent. They preferentially absorb water but reject salts at the root level, causing salts to build in the soil to the point where the plant's salt tolerance is surpassed.
2. Accumulators are plants that absorb salts and store them in plant cell vacuoles. Accumulators include glycophytes, salt-tolerant plants, and euhalophytes. Generally, these plants only have remedial value if they or sections are removed from the fields.
3. Conductor plants collect salts from the soil and carry the salt through the plant to the shoot surface, which is dispersed by wind. These plants can extract salt from the soil and disperse it broadly through wind dispersal.

Regarding the potential and advantages of halophytes, field experiments have shown that salt-tolerant plants may ameliorate salty soil, and Halophytes are considered the optimal choice for implementing sustainable and productive agriculture in highly saline soils. These plants offer numerous advantages, as they possess both economic and ecological value. Many halophytic species have a high economic significance, serving purposes such as food production, oil and medicine manufacturing, and vegetable cultivation. Additionally, they can be utilized as raw materials in the food processing industry, further enhancing their practical applications and economic potential [36]. Marine agriculture is a revolutionary modern agricultural technology that can be important for coastal regions. Considering that approximately 70% of the Earth's surface is covered by oceans, marine agriculture is poised to play a crucial role as a sustainable solution in the future. Certain halophytes, also referred to as saltwater vegetables, can be cultivated by irrigating them with seawater or planting them in saline and alkaline soils. These halophytes offer a unique advantage as they consist of edible parts that encompass both halophytes specifically adapted to saline conditions and traditional vegetables with varying degrees of salt tolerance. This makes them a promising option for meeting future food demands while utilizing oceanic resources sustainably [31].

Halophytes can be grown commercially to meet the demands of the biomass production sector regarding lignocellulosic matter and seed production with high oil content. Aside from the influence of direct saline water from the ocean, coastal regions endure saltwater intrusion in their groundwater supplies, and halophytes find such domains to be ideal for their growth. In environments where freshwater availability for irrigation is limited, it becomes a constraint for cultivating other crop varieties. However, halophytes are exceptionally well-suited to thrive in such conditions [71].

Compared to typical vegetables, seawater vegetables are regarded globally as natural, green, healthful, and organic meals high in protein, vitamins, and trace elements that people require, such as iron, selenium, and zinc. As a result, saltwater vegetables have gained popularity in recent years as perfect sustainable foods. They may be planted in salty soils or immediately along beaches, allowing them to not only remediate saline soil but also to prevent further coastal erosion. Plants that thrive in salt soils can also address the problem of food and fuel. Because of their salt resistance, such plants may grow beneath seawater, and many can be "cultivated and tamed" as food crops. *Salicornia, Sarcocornia* and *Chenopodium quinoa* are examples of salt-tolerant plants that become popular due to their nutritional value in food services [45, 1]. Due to their favorable crude protein (CP) content and satisfactory dietary composition, a diverse range of such plants can be utilized as fodder for various animals. Surprisingly, *Atriplex sp.* is regarded as a good feed source during food scarcity and drought.

From a medical perspective, various species of halophytes have been utilized in both traditional and modern medicine. Recent studies have demonstrated the effectiveness of plant extracts derived from these species in combating infections among humans, animals, and plants [13, 43]. Furthermore, pharmaceutical halophytes have gained significant importance in global healthcare (Yensen 2008; Husain et al. 2008; 107]. Numerous medicinal plants with halophytic properties are employed in the treatment of digestive, skin, respiratory, and genitourinary ailments, as well as pain, infections, fever, and toothache. These herbs, particularly in rural areas, are commonly used in anti-aging practices, where traditional medicine continues to be a valuable resource for managing minor illnesses (Ksouri et al. 2010, 2011, 2012) [43,81].

Aside from salt tolerance, halophytes may be resistant to various PTEs. Halophytes' metal resistance holds considerable promise for phytoremediation research. Halophytes have certain ways that boost their resistance to PTEs. Due to the frequent influx of pollutants, such as heavy metals, from industrial waste into saline regions, the utilization of phytoremediation to restore soils contaminated with both salinity and heavy metals has become an attractive prospect. This can be achieved through the use of salt-tolerant plants [26]. For instance, research conducted in various regions of Tunisia has reported instances of cadmium, lead, and nickel pollution [26]. Consequently, salt-tolerant plants have emerged as potential candidates for the removal or immobilization of heavy metals in saline soils, as indicated by several studies [58, 62]. The use of halophytes in the phytoremediation of soil salts is successful. *Atriplex triangularis, Suaeda glauca, Suaeda vera, Phragmites australis*, and are some of the halophytes employed in this technique [31].

3 Strategies for Remediation of Contaminated Environments

In order to deal with the issue of contaminated environments by f organic and inorganic pollutants, a huge effort has been deployed to develop innovative approaches to remediate those environments. Remediation is the process of using several methods to remove contaminants, including PTEs, from contaminated ecosystems, especially soil and water sources. The remediation methods have been classified into physical, chemical, and biological processes [102, 79, 3].

Various physicochemical and biological methods are used to remove environmental contaminants. However, the choice of the appropriate option depends on some factors like cost, effectiveness, commercial availability, and applicability to a wide spectrum of contaminants and potential negative side effects (toxicity) [2, 10, 89, 106]. Among physicochemical remediation methods are adsorption, electrokinetic technique, membrane filtration, granular activated carbon, photocatalysis, soil washing, chemical precipitation, floatation, ion exchange, coagulation, and flocculation [3].

For example, the adsorption procedure, which removes PTEs, implies complexation, chemisorption, adsorption-complexation on the surface and pores, micro precipitation, and ion exchange. This technique is an effective remediation process because of its various advantages including stability, utility, low cost, ease of operation, and performance [61]. Sani et al. (2017) reported a significant ability of granular activated carbon from African palm fruit to adsorb heavy metals like Cadmium, Copper, Nickel, and Lead from wastewater effluent. The electrokinetic method is intended to remove PTEs from soil. Electrokinetic remediation is a technique that involves the generation of an electrical field within a soil matrix. This is achieved by applying a low-voltage direct current to electrodes that are installed in the soil.

Consequently, PTEs may be mobilized, concentrated at the electrodes, and extracted from the soil [95]. Membrane technology is employed for the effective removal of hazardous pollutants from contaminated water. This method combines both membrane filtration and adsorption processes, allowing for the efficient elimination of various pollutants, including cationic heavy metals, anionic phosphates, and nitrates [40]. The method of soil washing offers a means to restore contaminated soils. In this process, known as in-situ soil washing, a washing solution is applied to the contaminated area either by flooding or sprinkling. The washing solution serves to extract contaminants from the soil, facilitating their removal and remediation. The mechanism of washing comprises desorption, chelation, dissolution, and other chemical actions under the addition of specific solutions [32]. This technique is advantageous for its flexibility of application, simplicity of operation high removal efficiency, and low cost [32]. Chemical precipitation technology is a widely used method for the remediation of groundwater, surface water, leachates, and industrial effluents from PTEs. This technique separates metal from solutions by forming metal hydroxides [18]. Flotation is a separation process that originated from mineral processing. This technique is mainly used in the field of water and wastewater treatment for heavy metal ions recovery. Ion exchange is another PTEs metals removal technique. The key mechanism in this technique is the exchangeability of anions between layered double hydroxides (LDHs). The aforementioned materials are referred to as ionic laminar composites. They consist of layers with a brucite-like structure that carry a positive charge. These layers are separated by an interlayer region containing charged compensating anions and solvation molecules. Due to their layered structure and large surface area, these composites have demonstrated significant potential for the remediation of heavy metals [33]. Coagulation and flocculation, combined with sedimentation and filtration, are also used to treat wastewater and effectively remove PTEs from wastewater [84]. Numerous coagulants are used in this process such as aluminum, ferrous sulfate, ferric chloride, and biochar [84]. Coagulation has the effect of destabilizing colloids by neutralizing the charges of particles. The techniques mentioned above have been reported to be more or less efficient for the removal of organic and inorganic contaminants, but the excessive use of chemical reagents may lead to the possibility of a secondary pollution issue. Bioremediation is one of the sustainable technologies that can be used to remove the harmful form of metals and contaminants to their less harmful state using biological mechanisms, including microorganisms or their enzymes and plant species, to restore the contaminated ecosystems [79].

4 Suitability and Potential of Biological Techniques (Phytoremediation)

Numerous biological techniques are used in the bioremediation process to restore the contaminated environment. The good example is the phytoremediation technique that involves the use of potential hyperaccumulating plants. The key mechanisms in this process are adsorption, sequestration, and uptake of potential trace metals, pesticides, and related emerging contaminants in both soil and aquatic environments [79]. The plant species used in this technique are characterized by a high tolerance for higher concentrations of toxic contaminants and a high capability of uptaking a considerable quantity of heavy metals from the contaminated soils and water [80]. Among these species are halophyte plants, considered hyperaccumulators in the aquatic environment and suitable for removing trace metals in contaminated soils, waterways, and wastewaters [79]. Halophytes have high salt tolerance and are suitable for phytoremediation by using different mechanisms such as exclusion, accumulation, and conduction [24]. These plant species grow under stress conditions such as flooding or doughty, excess light, and temperature and with the presence of toxic ions (Na^+ and Cl^-). They are also used as suitable as a part of the biosaline agriculture strategy for salt-affected lands for plant production thanks to their anatomical and physiological adaptations and rapid metabolic responses to promote osmotic adjustment [24]. For example, halophyte *Halogeton glomeratus* has been reported to be potentially effective in removing and accumulating salt and heavy metals from saline-contaminated soil [49]. Phytoremediation is considered a relatively cheap method thanks to its facile in-situ application compared to physical and chemical technologies (Fig. 2).

5 Mechanism of Halophytes' Adaptation in Polluted Environments

The cultivation and utilization of halophytes crops have emerged as a noteworthy trend in areas with limited agricultural suitability due to water scarcity. Halophytes offer an alternative solution for producing feeds, grains, grass forage, medicinal raw materials, and oil raw materials on fields irrigated with saline waters. Additionally, they contribute to the restoration of vegetation cover and enhancement of biological productivity in degraded pasturelands within arid regions. According to Korovin's research (1961), halophytes have been found to originate and thrive in regions characterized by significant soil salt concentrations, as identified across different geographic areas (Table 1). These were originally littorals, then alluvial plains, and finally secondary salt deposits in irrigated areas. The greatest concentration of halophyte species can be found in arid regions of Asia, Australia, America, and Africa because of these ecological conditions (Table 1).

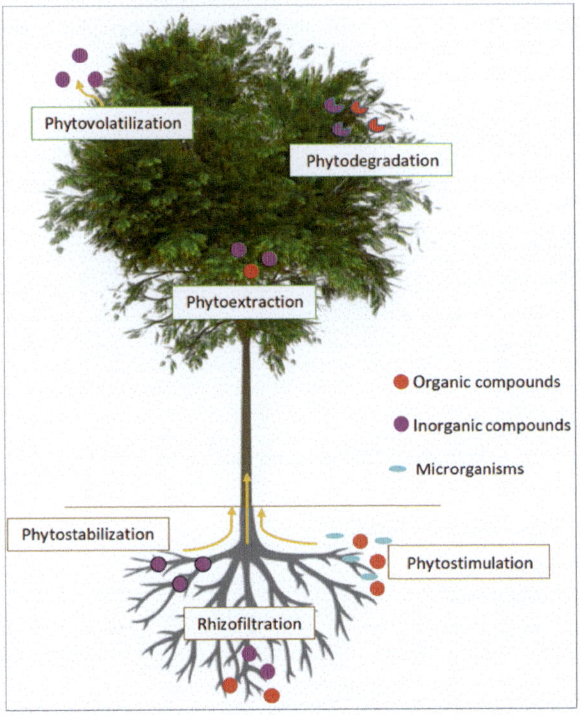

Fig. 2 Main plant mechanisms involved in phytoremediation to destroy, eliminate, transfer, or stabilize environmental pollutants (Garcia et al. 2020)

The *Chenopodiaceae Vent.*, *Poaceae Barnhart, Asteraceae Dumort., Plumbaginaceae Juss., Aizoaceae Rudolphi, Cyperaceae Juss., Papilionaceae Giseke, Tamaricaceae Link., Arecaceae Sch. Bip.*, and *Zygophyllaceae R. Br.* families contain the majority of halophytes. The families *Poaceae Barnhart* (137 species), *Asteraceae Dumort.* (69), *Plumbaginaceae Juss.* (57), and *Aizoaceae Rudolphi* (53) are not only exclusively halophytic in the flora, but also form the heart of halophytic plant communities in diverse parts of the world.

5.1 Adaptation of Halophytes in Arid and Semi-arid Regions

In arid and semi-arid regions, substantial areas experience prolonged periods of low-rate precipitation and are often devoid of surface hydrological resources such as lakes or rivers. Consequently, it is crucial to incorporate halophyte plants into local agriculture within these zones. Halophytes serve as alternative crops that not only contribute to the region's economic development but also play a vital role in freshwater conservation and social advancement. The combination of expansive saline-irrigated coastlines and the aridity of these regions creates a favorable environment for cultivating halophytes as alternative crops. Numerous studies have examined

Table 1 Distribution of halophyte species by geographical regions

Geographical distribution	Number of species	%
Asia	238	15.28
Australia and Asia	8	0.51
Africa	190	12.20
South America	148	9.50
Australia and Africa	5	0.32
Australia	219	14.06
Cosmopolite	39	2.50
North America	223	14.31
Pantropical	36	2.31
Central America	60	3.85
Sahara—Arabian and Sudanese regions	43	2.76
New World's tropics	14	0.90
Europe and North America	13	0.83
Europe	71	4.56
Mediterranean	70	4.49
Old World's tropics	67	4.30
America	29	1.86
Eurasia	15	0.96
Australia	31	1.99
Africa and America	4	0.26
Oceania	13	0.83
Australia and America	2	0.13
Africa and Asia	5	0.32
Boreal	2	0.13
Northern Hemisphere	1	0.06
Loss	**12**	**0.77**
Total	**1558**	**100**

the influence of climatic conditions, soil moisture content, and soil salinity on the development of halophytes. However, most of the existing research has focused on different halophyte species, including *S. europaea*, *S. stricta* Dumort, PuccineUia sp., Atriplex sp., and Distichlis sp. [47, 53, 59, 60, 69, 70, 77–93, 97]. Biotic and abiotic variables both influence biomass production and reproductive development in halophytes. Soil salinity, deposition and erosion of substrate by tidal action, depth of the water table, quantity of precipitation, temperature, and other microenvironmental factors can all have an impact on halophytes. Annuals' survival in salty environments is defined by their capacity to survive each growing season and reach the reproductive stage. In fact, all halophytes are adapted to environmental constraints, especially the climate, which controls their growth and geographic distribution in arid regions, explaining their distribution from coastal areas to mountain valleys (Khan, 2002). These constraints include moisture shortages caused by the osmotic and toxic effects of salts on plants as well as the physiological dryness of soil, which is preconditioned by a high ion content in the soil solution (Walter, 1976).

On the contrary, the rapid growth of industrialization has led to the substantial production and accumulation of potentially toxic elements (PTEs) like Cd, Cu, Pb, and Zn in tidal marsh soils [22, 30]. These elements, including Cd from pesticides and wood preservatives, Cr from tanneries and steel industries, Pb from battery manufacturing, herbicides, and petroleum production, and Hg from medical waste [5, 101], pose risks to human beings and other living organisms as they accumulate in the soil and water [39]. Addressing the pollution in these soils is essential but challenging, especially due to the high costs associated with standard remediation methods such as soil incineration or washing [85, 101]. The use of hyper-accumulating plants can significantly reduce these expenses. Certain halophytic plants (refer to Table 2) have the ability to thrive in contaminated environments and are commonly employed in phytoremediation techniques like phytoextraction and phytostabilization [52, 56].

The process of removing a PTE (e.g., Ag, Cd, Co, Mo, Pb, Zn) from soil by hyperaccumulating plants is referred to as phytoextraction. Plant roots absorb pollutants from the soil and store them in their aboveground sections throughout this process. The plant's biomass can then be harvested and disposed of. Species that may collect PTEs from sediments and accumulate them in their biomass include seagrasses and salt marsh plants [14, 56] Lewis and Devereux, 2009). Similar results have been observed using halophyte species such as *Sporobolus virginicus* [19], *Atriplex halimus* and *Tamarix smyrnensis* (Manousaki and Kalogerakis, 2011). It has been shown that *Atriplex halimus* can remove 4.15 kg of Cd and 2.2 kg of Zn per ha/year, assuming a plant output of 5 t DM/ha/year (Lutts et al. 2004). Lokhande et al. (2011) looked at *Sesuvium portulacastrum* as a viable option for reclaiming As-affected land since it had strong plant growth and the ability to acquire up to 155 g/g dry weight of As. However, the majority of known hyperaccumulators are slow growing, limiting their ability to extract pollutants from the soil. According to a previous study (El Amier et al. 2017) about the bioaccumulation and translocation of PTEs from coastal soil by wild halophytes, six natural halophytes were gathered from the soil of the northern Nile Delta in order to assess their phytoremediation

Table 2 Selected examples of PTEs accumulation (mg/kg DM) by halophyte plants

Plant species	PTEs	Amount accumulated (mg/kg DM)	References
Sesuvium portulacastrum	As	155	Lokhande et al. (2011)
Atriplex halimus	Cd	830	Lutts et al. (2004)
Atriplex halimus	Zn	440	Lutts et al. (2004)
Atriplex species	B	192	Watson et al. [96]
Sesuvium portulacastrum	Cd	350–700	Ghnaya et al. [25]

Table 3 Mean concentration (μg g^{-1} dry weight) of PTEs ± standard deviation in tissues of studied halophytes (2017)

Plant species	Plant part	Fe	Pb	Ni	Co	Cd
Atriplex halimus	Shoot	283.5 ± 28.4	4.35 ± 0.4	1.5 ± 0.2	0.8 ± 0.1	1.8 ± 0.2
	Root	458.69 ± 45.9	3.2 ± 0.3	2.67 ± 0.3	1.48 ± 0.1	2.4 ± 0.2
Limoniastrum monopetalum	Shoot	290.1 ± 29.1	4.3 ± 0.4	1.7 ± 0.2	0.9 ± 0.1	0.6 ± 0.1
	Root	311.45 ± 31.1	2.48 ± 0.2	2.89 ± 0.3	1.24 ± 0.1	0.9 ± 0.1
Limonium pruinosum	Shoot	360.4 ± 36.0	5.0 ± 0.5	1.9 ± 0.2	1.0 ± 0.1	0.8 ± 0.1
	Root	476.2 ± 47.6	3.15 ± 0.3	3.42 ± 0.3	1.56 ± 0.2	1.78 ± 0.2
Suaeda maritima	Shoot	234.36 ± 23.4	2.22 ± 0.2	0.74 ± 0.1	0.5 ± 0.1	0.81 ± 0.1
	Root	695.3 ± 69.5	1.56 ± 0.2	2.59 ± 0.3	1.12 ± 0.2	1.41 ± 0.1
Suaeda pruinosa	Shoot	186.89 ± 18.7	5.11 ± 0.5	1.59 ± 0.2	0.8 ± 0.1	0.82 ± 0.1
	Root	636.14 ± 63.6	3.23 ± 0.3	3.58 ± 0.4	1.9 ± 0.1	2.34 ± 0.2
Zygophyllum aegyptium	Shoot	238.45 ± 23.8	4.09 ± 0.4	1.54 ± 0.2	0.85 ± 0.1	0.08 ± 0.0
	Root	90.98 ± 9.1	5.76 ± 0.6	2.21 ± 0.2	1.11 ± 0.1	0.13 ± 0.0

potential of PTEs. To assess their potential for phytoremediation in contaminated soils, the soil samples, aboveground plant parts, and roots were analyzed for the total concentrations of Fe, Pb, Ni, Co, and Cd using an atomic absorption spectrometer. The results of the analysis suggest that this particular type of plant holds promise for effectively remediating polluted soils.

Plants are an excellent instrument for phytoremediation. The natural capacity of certain plants to clean up tainted soil, air, and water is referred to as hyper-accumulation (Reichenauer and Germida, 2008). During the research periods, authors measured the PTEs concentrations in various tissues of numerous plant species occurring naturally along the Deltaic Mediterranean coast (Table 3). PTEs concentrations in plant species in the research region were in the following order: Fe > Pb > Ni > Cd > Co.

5.2 *Adaptation of the Halophyte in Saline Soils*

Salinity now affects approximately 20% of the world's agricultural fields and half of the world's irrigated lands (Rhoades and Loveday, 1990). According to another report, salinity affects roughly 400 million/hectare of land [4]. If plant salt tolerance is not improved, significant portions of the earth will remain uncultivated. However, developing a salt-tolerant plant could take several years. It is because genes controlled by salt stress can be discovered using either RNA [37] or proteins [72], and finding a critical gene for salinity tolerance remains a pipe dream. There is more than one mechanism working in salt-tolerant plants against soil salinity, so it is critical to research the mechanism functioning at each level in full detail in order to build a complete understanding of salt stress and use this knowledge to improve crop resistance to salt stress. Under such soil conditions, halophytes can be highly valuable, they can be exploited for industrial, ecological, and agricultural applications (Koyro et al. 2011). Salinity exerts multiple effects on various aspects of plant physiology simultaneously. These effects include a decrease in water absorption by the plants (Epstein, 1980), a rapid decline in growth rates, and the activation of metabolic changes similar to those triggered by water stress (Epstein, 1980). The presence of halophytes in areas with relatively high salt concentrations indicates one of the following: a necessity for higher salt concentrations, a tolerance for excessive salt levels, or a diminished competitive ability compared to less stress-tolerant plants in the environment [94]. As observed globally and domestically, halophyte farming offers solutions to three main challenges: the preservation of freshwater resources and financial considerations, environmental reclamation (including soil fertility), and agricultural production encompassing grains, grass forage, feeds, and oil and medicinal raw materials (O'Leary, 1985). The cultivation of halophyte crops can primarily take place on sandy soils in desert environments, utilizing irrigation with saline water from sources such as the sea, groundwater, or drainage. According to Glenn and Watson [29], approximately 15% of uncultivated lands in coastal and continental deserts could become viable for farming if irrigated with seawater, thereby providing potential for cultivating new crops on an estimated 130 million hectares. Steiner [86] categorized halophytes based on their reaction to internal salt content. According to him, salt marsh halophytes are classified into two types: salt regulating and salt accumulating. Halophytes are also classed as excluders vs includes based on their internal salt concentration [9]. All halophytes are classified into numerous ecological groupings based on substrate salinity and soil texture. According to [8], hyperhalophytes account for 45.25% of the total,xerohalophytes account for 29.54%,psammohalophytes account for 16.29%,and other categories (chasmophytes, weeds, and phreatophytes) account for 6.35%. Indeed, Halophytes have habitat-forming and habitat-optimization capabilities, which result in a beneficial effect on salinized soils. Halophyte plants enable a significant reduction in physical evaporation, a lower groundwater level, salt carryover by the top mass, and soil desalinization due to efficient shading of the soil surface by the top mass, pumping functions, and biological-drainage functions (Shamsutdinov, 2005, 2002). The incorporation of fresh organic matter from plants enhances the physical, chemical, and biological properties of the soil, while also influencing factors such as pH, electrical conductivity, and hydraulic conductivity. Based on domestic experience, the key principle of establishing a meliorative crop rotation involves the initial utilization of halophytes in the early years, followed by a transition to mixed plantings of a halophyte along with a feed crop. As the soil gradually desalinizes, the area dedicated to the feed crop is expanded. Once the soil has been completely desalinized, a pure feed crop is cultivated (Shamsutdinov, 2005). For medium saline soils and highly saline medium-loam soils with chloride salinization levels not exceeding 0.6%, the biological approach to soil desalinization is recommended as an effective method.

6 Limitations of Phytoremediation Using Halophytes

Nonetheless, phytoremediation needs more work than just growing greenery and believing that the contamination would disappear with minimum upkeep. It necessitates an awareness of the processes that must occur, the plants chosen, and what must be done to promote plant development [91]. Given the large number of candidates, only a small number of plants have been studied,another limitation is a lack of understanding of basic plant remediation mechanisms; thus, extensive research is required to identify suitable plant species that can colonize the polluted area and remove, degrade, or immobilize the contaminant of interest [57].

As previously mentioned, phytoremediation is a low-cost, low-maintenance, environmentally benign, and renewable resource for polluted environments. In some cases, such as pesticide-contaminated soil or groundwater, phytoremediation may be the only viable and cost-effective in-situ method for removing harmful chemicals. However, organic PTEs pollutant absorption and translocation from soil via the plant root may be limited to chemicals that are not very hydrophobic, and through leaves to compounds that are not hydrophilic [11]. Furthermore, in instances where modest doses of PTEs were extensively spread in the environment, the technique would most likely have minimal influence. Even in the presence of high levels of PTEs in a specific environmental, such as soil or water, phytoremediation may be helpful as a long-term solution. This is because plant development is affected by various environmental and climatic conditions, such as water and nutrient availability, soil type and pH, and so on. Even under ideal conditions, plant growth, and PTEs removal/degradation may not be efficient 100%. These constraints need the combination of phytoremediation with other more rapid clean-up alternatives and a more cost-effective use of the biomass generated for non-food applications, such as electricity generation [64].

On the other hand, the importance of vegetation and plant biomass as an absorbent and long-term sink for dispersed PTEs should not be overlooked. There is also a need to identify techniques to improve pollutant PTEs absorption and degradation in plants. Because microbial activity in the rhizosphere is known to help release pollutants residues in soil, which can improve absorption and transformation by plants, a combination of microbial bioremediation and phytoremediation is more likely to be successful. While advances in biotechnology sciences have transformed many fields of conventional research, they have had little influence on phytoremediation until now. The quantity of information developed in plant biotechnology in recent years has been vast, and many more untapped routes may be pursued to create or improve a specific pathway for the phytoremediation of PTEs in the environment [15].

7 Future Perspectives and Conclusions

Increasing industrial and agricultural production to support global population growth makes it necessary to manage pollutant, persistent, and harmful wastes to the whole trophic network. Therefore, PTEs contaminated area's remediation is an essential practice to preserve the health of ecosystems. To this end, in this chapter, several strategies were mentioned; moreover, the need to choose the most appropriate method using economic and environmental sustainability criteria was emphasized. In fact, physical and chemical methods can be expensive and generate secondary pollution issues. Otherwise, the use of biological techniques can be a feasible strategy for the remediation of contaminated environments. Indeed, pollutants removal through phytoextraction by hyperaccumulating plants such as halophytes not only can reduce the cost of cleaning contaminated soils, but also allows for better management of salinity-compromised areas. Furthermore, they will undoubtedly be able to provide a habitat for species in areas where pollutants seriously damaged biodiversity. In the future, because of their disparate other uses, from food consumption to manufacturing use, they may increasingly be a key element for food security in sustainable production, especially in marginal or degraded areas. In conclusion, all the aforementioned advantages should provide an incentive for new studies in order to fill the lack of knowledge about, for example, basic plant remediation mechanisms. Finally, identifying methodologies that can enable optimization of phytoremediation processes, e.g., by applying them in conjunction with microbial bioremediation, is essential.

References

1. Fita, A., Rodríguez-Burruezo, A., Boscaiu, M., Prohens, J., Vicente, O.: Breeding and domesticating crops adapted to drought and salinity: a new paradigm for increasing food production. Front. Plant Sci. **6** 978 (2015)
2. Abbar, B., Alem, A., Marcotte, S., Pantet, A., Ahfir, N.-D., Bizet, L., Duriatti, D.: Experimental investigation on removal of heavy metals (Cu2+, Pb2+, and Zn2+) from aqueous solution by flax fibres. Process. Saf. Environ. Prot. **109**, 639–647 (2017)
3. Akhtar, F.Z., Archana, K.M., Krishnaswamy, V.G. and Rajagopal, R.: Remediation of heavy metals (Cr, Zn) using physical, chemical and biological methods: a novel approach. SN Appl. Sci. **2**, 1–14 (2020)
4. Al-Sadi, A.M., Al-Masoudi, R.S., Al-Habsi, N., Al-Said, F.A., Al-Rawahy, S.A., Ahmed, M., Deadman, M.L.: Effect of salinity on Pythium damping-off of cucumber and on the tolerance of Pythium aphanidermatum. Plant Pathol. **59**, 112–120 (2010)
5. Ali, H., Khan, E., Sajad, M.A.: Phytoremediation of heavy metals—concepts and applications. Chemosphere **91**, 869–881 (2013)
6. Anjum, N.A., Duarte, B., Caçador, I., Sleimi, N., Duarte, A.C., Pereira, E.: Biophysical and biochemical markers of metal/metalloid-impacts in salt marsh halophytes and their implications. Front. Environ. Sci. **4**,(2016). https://doi.org/10.3389/fenvs.2016.00024
7. Antoniadis, V., Levizou, E., Shaheen, S.M., Ok, Y.S., Sebastian, A., Baum, C., Prasad, M.N.V., Wenzel, W.W., Rinklebe, J.: Trace elements in the soil-plant interface: phytoavailability, translocation, and phytoremediation–a review. Earth Sci. Rev. **171**, 621–645 (2017). https://doi.org/10.1016/j.earscirev.2017.06.005

8. Aronson, J.: Haloph. Office of Arid Lands Studies, Univ. Arizona Tucson, Arizona, A database of salt tolerant plant of the world (1989)
9. Ashraf, M., Hameed, M., Arshad, M., Ashraf, Y., Akhtar, K.: Salt tolerance of some potential forage grasses from Cholistan desert of Pakistan. In: M. A. Khan and D.J. Weber (eds.). Ecophysiol, high salinity tolerance plants. Springer, Netherlands (2006), pp. 31–54
10. Aziz, H., Wang, X., Murtaza, G., Ashar, A., Hussain, S., Abid, M., Murtaza, B., Saleem, M.H., Fiaz, S., Ali, S.: Evaluation of compost and biochar to mitigate chlorpyrifos pollution in soil and their effect on soil enzyme dynamics. Sustainability **13**, 9695 (2021)
11. Babu, S.M.O.F., Hossain, M.B., Rahman, M.S., Rahman, M., Ahmed, A.S.S., Hasan, M.M., Rakib, A., Emran, T.B., Xiao, J., Simal-Gandara, J.: Phytoremediation of toxic metals: a sustainable green solution for clean environment. Appl. Sci. **11**, 10348 (2021). https://doi.org/10.3390/app112110348
12. Bandara, T., Franks, A., Xu., Jianming, Bolan, N., Wang, H., Tang, C.: Chemical and biological immobilization mechanisms of potentially toxic elements in biochar-amended soils. Crit. Rev. Environ. Sci. Technol. (2019). https://doi.org/10.1080/10643389.2019.1642832
13. Breckle, S.W.: Vegetation, climate and soil: 50 years of global ecology. Prog. Bot. **80**, 1–63 (2019)
14. Cambrolle, J., Redondo-Gomez, S., Mateos-Naranjo, E., Figueroa, M.E.: Comparison of the role of two Spartina species in terms of phytostabilization and bioaccumulation of metals in the estuarine sediment. Mar. Pollut. Bull. **56**, 2037–2042 (2008)
15. Chaudhry, Q., Schröder, P., Werck-Reichhart, D., Grajek, W., Marecik, R.: Prospects and limitations of phytoremediation for the removal of persistent pesticides in the environment. Environ. Sci. Pollut. Res. Int. **9**(1), 4–17 (2002). https://doi.org/10.1007/BF02987313. PMID: 11885417
16. Christofilopoulos, S., Syranidou, E., Gkavrou, G., Manousaki, E., & Kalogerakis, N.: The role of halophyte *Juncus acutus* L. in the remediation of mixed contamination in a hydroponic greenhouse experiment: Remediation of mixed contamination by *J. acutus* L. *Journal of Chemical Technology & Biotechnology*, *91*(6), 1665–1674 (2016). https://doi.org/10.1002/jctb.4939
17. Dhaliwal, S.S., Singh, J., Taneja, P.K., Mandal, A.: Remediation techniques for removal of PTEs from the soil contaminated through different sources: a review. Environ. Sci. Pollut. Res. **27**(2), 1319–1333 (2020). https://doi.org/10.1007/s11356-019-06967-1
18. Djedidi, Z., Bouda, M., Souissi, M.A., Cheikh, R., Mercier, G., Tyagi, R.D., Blais, J.F. Metals removal from soil, fly ash and sewage sludge leachates by precipitation and dewatering properties of the generated sludge. J. Haz. Mat. 172(2):1372–1382 (2009). https://doi.org/10.1016/j.jhazm at.2009.07.144
19. Eid, M.A., Eisa, S.S.: The use of some halophytic plants to reduce Zn, Cu and Ni in soil. Aust. J. Basic Appl. Sci. **4**(7), 1590–1596 (2010)
20. El-Amier, Y.A., Alghanem, S.M., Alzuaibr, F.M.: Bioaccumulation and translocation of heavy metals from coastal soil by wild halophytes. Am. J. Environ. Prot. **5**(2), 52–60 (2017). https://doi.org/10.12691/env-5-2-4
21. Epstein, E.: Responses of plants to saline environments. In: Rains, D.W., Valentine, R.C., Hollaender, A. (eds.) Genetic engineering of osmoregulation, pp. 7–21. Plenum Press, New York (1980)
22. Flowers, T.J., Galal, H.K., Bromham, L.: Evolution of halophytes: multiple origins of salt tolerance in land plant. Funct. Plant Biol. **37**, 604–612 (2010)
23. Flowers, T. J., & Muscolo, A.: Introduction to the Special Issue: Halophytes in a changing world. *AoB PLANTS*, *7* (2015). https://doi.org/10.1093/aobpla/plv020
24. Garcia, J.S., França, M.G.C., Prasad, M.N.V.: (2020). Haloremediation for Amelioration of Salinity. In: Grigore, MN. (eds) Handbook of Halophytes. Springer, Cham. https://doi.org/10.1007/978-3-030-17854-3_88-1
25. Ghnaya, T., Nouairi, I., Slama, I., Messedi, D., Grignon, C., Adbelly, C., Ghorbel, M.H.: Cadmium effects on growth and mineral nutrition of two halophytes: *Sesuvium portulacastrum* and *Mesembryanthemum crystallinum*. J. Plant Physiol. **162**, 1133–1140 (2005)

26. Ghnaya, T., Slama, I., Messedi, D., Grignon, C., Ghorbel, M.H. and Abdelly, C.: Cd-induced growth reduction in the halophyte Sesuvium portulacastrum is significantly improved by NaCl. J. Plant Res. **120**, 309–316 (2007)
27. Ghori, N.-H., Ghori, T., Hayat, M.Q., Imadi, S.R., Gul, A., Altay, V., Ozturk, M.: Heavy metal stress and responses in plants. Int. J. Environ. Sci. Technol. **16**(3), 1807–1828 (2019). https://doi.org/10.1007/s13762-019-02215-8
28. Glenn, E., Miyamoto, S., Moore, D., Brown, J. J., Thompson, T. L., & Brown, P.: Water requirements for cultivating Salicornia bigelovii Torr. With seawater on sand in a coastal desert environment. *Journal of Arid Environments, 36*(4), 711–730 (1997). https://doi.org/10.1006/jare.1997.0253
29. Glenn, E.P. and Watson, M.C.: Halophyte crops for direct seawater irrigation. In: H. Lieth and A.A. Al Masoom (Editors), Towards the Rational Use of High Salinity Tolerant Plants. vol. 1. Kluwer Academic, Dordrecht, pp 379–386
30. Govindasamy, C., Arulpriya, M., Ruban, P., Francisca, J.L., Ilayaraja, A.: Concentration of heavy metals in seagrasses tissue of the Palk Strait, Bay of Bengal. Int. J. Environ. Sci. **2**, 145–153 (2011)
31. Grigore, M. N., Toma, C.: Morphological and anatomical adaptations of halophytes: A review. In: Grigore, M. N. (eds) Handbook of Halophytes. Springer, Cham (2021). https://doi.org/10.1007/978-3-030-57635-6_37
32. Gu, F., Zhang, J., Shen, Z., Li, Y., Ji, R., Li, W., Zhang, L., & Han, J.: (2022) A review for recent advances on soil washing remediation technologies
33. Guan, X., Yuan, X., Zhao, Y., Wang, H., Wang, H., Bai, J., Li, Y.: Application of functionalized layered double hydroxides for heavy metal removal: a review. Sci. Total. Environ. **838**, 155693 (2022). https://doi.org/10.1016/j.scitotenv.2022.155693
34. H. Walter, Die Vegetation der Erde in Öko-physiologischer Betrachtung, vol. 2: Die gemäßigten und arktischen Zonen (G. Fischer, Stuttgart, 1968; Progress, Moscow, 1974)
35. Hasanuzzaman, M., Nahar, K., Alam, Md.M., Bhowmik, P.C., Hossain, Md.A., Rahman, M.M., Prasad, M.N.V., Ozturk, M., Fujita, M.: Potential use of halophytes to remediate saline soils. Biomed. Res. Int. **2014**, 1–12 (2014). https://doi.org/10.1155/2014/589341
36. Hasnain, M., Abideen, Z., Ali, F., Hasanuzzaman, M., El-Keblawy, A.: Potential of halophytes as sustainable fodder production by using saline resources: a review of current knowledge and future directions. Plants **12**, 2150 (2023). https://doi.org/10.3390/plants12112150
37. Kawasaki, S., Borchert, C., Deyholos, M., Wang, H., Brazille, S., Kawai, K., Galbraithand, D., Bohnert, H.J.: Gene expression profiles during the initial phase of salt stress in rice. Plant Cell **13**, 889–905 (2001)
38. Khan, M.A., Gul, B., Weber, D.J.: Effect of temperature, and salinity on the germination of *Sarcobatus vermiculatus*. Biol. Plant. **45**, 133–135 (2002)
39. Khan, S., Hesham, A.E.L., Qiao, M., Rehman, S., He, J.Z.: Effects of Cd and Pb on soil microbial community structure and activities. Environ. Sci. Pollut. Res. **17**, 288–296 (2010)
40. Khulbe, K.C., Matsuura, T.: Removal of heavy metals and pollutants by membrane adsorption Techniques. Appl. Water Sci. 8(1):19–49 (2018). https://doi.org/10.1007/s13201-018-0661-6
41. Korovin, E. P.: Vegetation of central Asia and Southern Kazakhstan (Izd. Akad. Nauk UzbSSR, Tashkent, 1961), vol. 1 [in Russian]
42. Koyro, H.-W., Khan, M.A., Lieth, H.: Halophytic crops: a resource for the futureto reduce the water crisis? Emirates J. Food Agric. **23**(1), 01–16 (2017). https://doi.org/10.9755/ejfa.v23i1.5308
43. Ksouri, R., Ksouri, W.M., Jallali, I., Debez, A., Magné, C., Hiroko, I. and Abdelly, C.: Medicinal halophytes: potent source of health promoting biomolecules with medical, nutraceutical and food applications. Critical Rev. Biotech. **32**(4), 289–326 (2012). https://doi.org/10.3109/07388551.2011.630647
44. Kyzas, G.Z., Matis, K.A.: Flotation in water and wastewater treatment. Processes 6(8), 2018. https://doi.org/10.3390/pr6080116
45. Ladeiro, B.: Saline agriculture in the 21st Century: using salt contaminated resources to cope food requirements. J. Bot. 1–7. https://doi.org/10.1155/2012/310705

46. Lamouri, S., et al.: Control of the γ-alumina to α-alumina phase transformation for an optimized alumina densification. Boletin de la Sociedad Espanola de Ceramica y Vidrio 56(2), 47–54 (2017). https://doi.org/10.1016/j.bsecv.2016.10.001
47. Langlois, J.: Cultures san sol de Salicornia stricta Dumort. Revue G~ndral de Botanique 74, 176–196 (1967)
48. Lewis, M.A., Devereux, R.: Non-nutrient anthropogenic chemicals in seagrass ecosystems: fate and effects. Environ. Toxicol. Chem. **28**, 644–661 (2009). https://doi.org/10.1897/08-201.1
49. Li, Y., Chen, Y., Wu, J.: Enhancement of methane production in anaerobic digestion process: a review. Appl. Energy **240**, 120–137 (2019). https://doi.org/10.1016/j.apenergy.2019.01.243
50. Li, B., Wang, J., Yao, L., et al.: Halophyte *Halogeton glomeratus*, a promising candidate for phytoremediation of heavy metal-contaminated saline soils. Plant Soil **442**, 323–331 (2019). https://doi.org/10.1007/s11104-019-04152-4
51. Lokhande, V.H., Srivastava, S., Patade, V.Y., Dwivedi, S., Tripathi, R.D., Nikam, T.D., Suprasanna, P.: Investigation of arsenic accumulation and tolerance in *Sesuvium portulacastrum* (L.). 902. Chemosphere 82, 529–534 (2011)
52. Lutts, S., Lefe'vre, I., Delpe´re´e, C., Kivits, S., Dechamps, C., Robledo, A., Correal, E.,: Heavy metal accumulation by the halophyte species *Mediterranean saltbush*. J. Environ. Qual. **33**, 1271–1279 (2004)
53. Macke, A., Ungar, I.A.: The effect of salinity on seed germination and early growth of *Puccinellia nuttaliana*. Can. J. Bot. **49**, 515–520 (1971)
54. Mahar, A., Wang, P., Ali, A., Awasthi, M.K., Lahori, A.H., Wang, Q., Li, R., Zhang, Z.: Challenges and opportunities in the phytoremediation of PTEs contaminated soils: a review. Ecotoxicol. Environ. Saf. **126**, 111–121 (2016). https://doi.org/10.1016/j.ecoenv.2015.12.023
55. Manousaki, E., Kalogerakis, N.: Halophytes—an emerging trend in phytoremediation. Int. J. Phytorem. **13**(10), 959–969 (2011). https://doi.org/10.1080/15226514.2010.532241
56. Manousaki, E., Kalogerakis, N.: Halophytes present new opportunities in phytoremediation of PTEs and saline soils. Ind. Eng. Chem. Res. **50**(2), 656–660 (2011). https://doi.org/10.1021/ie100270x
57. Manousaki, E., Kalogerakis, N.: Halophytes present new opportunities in phytoremediation of heavy metals and saline soils. Ind. Eng. Chem. Res. **50**, 656–660 (2011)
58. Manousaki, E., Kalogerakis, N. Phytoextraction of Pb and Cd by the Mediterranean saltbush (*Atriplex halimus L.*): metal uptake in relation to salinity. Environ. Sci. Pollut. Res. **16**, 844–854 (2009). https://doi.org/10.1007/s11356-009-0224-3
59. McGraw, D.C., Ungar, I.A.: Growth and survival of the halophyte *Salicornia europaea* L. under saline field conditions. Ohio J. Sci. **81**, 109–113 (1981)
60. McMahon, K.A., Ungar, I.A.: Phenology, distribution and survival of *Atriplex triangularis* Willd. in an Ohio salt pan. Am. Midl. Nat. **100**, 1–14 (1978)
61. Muharrem, I., Olcay, K. I.: An overview of adsorption technique for heavy metal removal from water/wastewater: a critical review. Int. J. Pure Appl. Sci. 3(2):10–19 (2017). https://doi.org/10.29132/ijpas.372335
62. Nedjimi, B. and Daoud, Y.: Ameliorative effect of CaCl2 on growth, membrane permeability and nutrient uptake in Atriplex halimus subsp. schweinfurthii grown at high (NaCl) salinity. Desalination **249**(1), 163–166 (2009)
63. Nguyen, K.A., Liou, Y.A., Tran, H.P., et al.: Soil salinity assessment by using near-infrared channel and Vegetation Soil Salinity Index derived from Landsat 8 OLI data: a case study in the Tra Vinh Province, Mekong Delta, Vietnam. Prog Earth Planet Sci **7**, 1 (2020). https://doi.org/10.1186/s40645-019-0311-0
64. Oladoye, P. O., Olowe, O. M., Asemoloye, M. D.: Phytoremediation technology and food security impacts of heavy metal contaminated soils: a review of literature. Chemosphere *288*(Part 2), 132555 (2022), ISSN 0045-6535, https://doi.org/10.1016/j.chemosphere.2021.132555
65. Ozturk, A., Yarci, C., & Ozyigit, I. I.: Assessment of heavy metal pollution in Istanbul using plant (*Celtis australis* L.) and soil assays. Biotechnol. Biotechnol. Equip. **31**(5), 948–954 (2017). https://doi.org/10.1080/13102818.2017.1353922

66. Panta, S., Flowers, T., Lane, P., Doyle, R., Haros, G., Shabala, S.: Halophyte agriculture: success stories. Environ. Exp. Bot. **107**, 71–83 (2014). https://doi.org/10.1016/j.envexpbot.2014.05.006
67. Qasim, M., Irfan, A., Munaweer, R., Bilquees, G., & Ajmal, M. K.: Effect of extraction solvents on polyphenols and antioxidant activity of medicinal halophytes. Pak. J. Bot. *48*(2) (2016)
68. Reichenauer, T.G., Germida, J.J.: Phytoremediation of organic contaminants in soil and groundwater. Chemsuschem **1**(8–9), 708–717 (2008)
69. Riehl, T.E., Ungar, I.A.: Growth and ion accumulation in *Salicornia europaea* under saline field conditions. Oecologia **54**, 193–199 (1982)
70. Rivers, W.G., Weber, D.J.: The influence of salinity and temperature on seed germination in *Salicornia bigelovii*. Physiol. Plant. **24**, 73–75 (1971)
71. Rozema, J., Schat, H.: Salt tolerance of halophytes, research questions reviewed in the perspective of saline agriculture. Environ. Exp. Bot. **92** 83–95 (2013)
72. Salekdeh, G.H., Siopongco, J., Wade, L.J., Ghareyazie, B., Bennett, J.: Aproteomic approach to analyzing drought and salt responsiveness in rice. Field Crops Res. **76**, 199–219 (2002)
73. Sani, A., Hussaini, K., Sani, H.M.: Evaluation of removal efficiency of heavy metals by low-cost activated carbon prepared from African palm fruit. Appl Water Sci *7*(6), 3151–3155 (2017). https://doi.org/10.1007/s13201-016-0460-x
74. Santos, E.S., Abreu, M.M., Peres, S., Magalhães, M.C.F., Leitão, S., Pereira, A.S., Cerejeira, M.J.: Potential of *Tamarix africana* and other halophyte species for phytostabilisation of contaminated salt marsh soils. J. Soils Sediments **17**(5), 1459–1473 (2017). https://doi.org/10.1007/s11368-015-1333-x
75. Santos, J., Al-Azzawi, M., Aronson, J., Flowers, T.J.: EHALOPH a database of salt-tolerant plants: helping put halophytes to work. Plant Cell Physiol. **57**(1), e10–e10 (2016). https://doi.org/10.1093/pcp/pcv155
76. Santos, M. S. S., Pedro, C. A., Gonçalves, S. C., & Ferreira, S. M. F.: Phytoremediation of cadmium by the facultative halophyte plant *Bolboschoenus maritimus* (L.) Palla, at different salinities. Environ. Sci. Pollut. Res. *22*(20), 15598–15609 (2015). https://doi.org/10.1007/s11356-015-4750-x
77. Santos, M. S. S., Pedro, C. A., Gonçalves, S. C., & Ferreira, S. M. F. (2015). Phytoremediation of cadmium by the facultative halophyte plant *Bolboschoenus maritimus* (L.) Palla, at different salinities. Environ. Sci. Pollut. Res., *22*(20), 15598–15609. https://doi.org/10.1007/s11356-015-4750-x
78. Santos, M. S. S., Pedro, C. A., Gonçalves, S. C., & Ferreira, S. M. F. (2015). Phytoremediation of cadmium by the facultative halophyte plant Bolboschoenus maritimus (L.) Palla, at different salinities. *Environmental Science and Pollution Research*, *22*(20), 15598–15609. https://doi.org/10.1007/s11356-015-4750-x
79. Sarker, A., Al, A., Mohammad, D., Das, K., Nandi, R., Waheda, M., Ansary, R., Reza, A., Islam, T., Islam, T.: Chemosphere Biological and green remediation of heavy metal contaminated water and soils: a state-of-the-art review. Chemosphere **332**(May), 138861 (2023). https://doi.org/10.1016/j.chemosphere.2023.138861
80. Sarwar, N., Imran, M., Shaheen, M.R., Ishaque, W., Kamran, M.A., Matloob, A., Rehim, A., Hussain, S.: Phytoremediation strategies for soils contaminated with heavy metals: modifications and future perspectives. Chemosphere **171**, 710–721 (2017). https://doi.org/10.1016/j.chemosphere.2016.12.116
81. Savithramma, N., Sulochana, C. and Rao, K.N.: Ethnobotanical survey of plants used to treat asthma in Andhra Pradesh, India. J. Ethnopharmacology **113**(1), 54–61 (2007)
82. Sh, Z., Shamsutdinov and N. Z. Shamsutdinov,: Halophytic crop farming. Sovetskii Sport, Moscow (2005). [in Russian]
83. Shamsutdinov, Z. Sh., Shamsutdinov, N. Z.: Biogeocenotic principles and methods of degraded pastures phytomelioration in Central Asia and Russia. In: Ahmad, R., Malik, K. A., Prospects for saline agriculture, (Springer, 2002), pp. 29–35

84. Shao, F., Xu, J., Chen, F., Liu, D., Zhao, C., Cheng, X., Zhang, J.: Insights into olation reaction-driven coagulation and adsorption: a pathway for exploiting the surface properties of biochar. Sci. Total Environ. 854, 158595 (2023). https://doi.org/10.1016/j.scitotenv.2022.158595
85. Sheoran, V., Sheoran, A., Poonia, P.: Role of hyperaccumulators in phytoextraction of metals from contaminated mining sites: a review. Crit. Rev. Environ. Sci. Technol. **41**, 168–214 (2011)
86. Steiner, M.: To the ecology of the salt march of the Nordostlichen united countries of Nordamerika. Jahrb. Know. Offered **81**, 94 (1934)
87. Sumiahadi, A., Acar, R.: A review of phytoremediation technology: PTEs uptake by plants. IOP Conf. Ser.: Earth Environ. Sci. **142**, 012023 (2018). https://doi.org/10.1088/1755-1315/142/1/012023
88. Szabolcs, I.: Salt affected soils as the ecosystem for halophytes. In: Squires, V.R., Ayoub, A.T. (eds), Halophytes as a resource for livestock and for rehabilitation of degraded lands. Tasks for Vegetation Science, vol 32. Springer, Dordrecht. https://doi.org/10.1007/978-94-011-0818-8_3
89. Tariq, F., Wang, X., Saleem, M.H., Khan, Z.I., Ahmad, K., Saleem Malik, I., Munir, M., Mahpara, S., Mehmood, N., Ahmad, T., Memona, H., Ugulu, I., Fiaz, S., Ali, S.: Risk assessment of heavy metals in basmati rice: implications for public health. Sustainability **13**, 8513 (2021)
90. Tiku, B.L.: Effect of salinity on the photosynthesis of the halophyte *Salicornia rubra* and *Distichlis stricta*. Physiol. Plant. **37**, 23–28 (1976)
91. USEPA. EPA/water quality, environmental matters. environmental protection agency, United States, 76 (2001)
92. Ungar, I.A.: Population characteristics, growth, and survival of the halophyte *Salicornia europaea*. Ecology **68**, 569–575 (1987)
93. Ungar, I.A.: Population ecology of halophyte seeds. Bot. Rev. **53**, 301–334 (1987)
94. Ungar, I.A.: Ecophysiology of vascular halophytes. CRC Press, Boca Raton (1991)
95. Venkata Sivapullaiah, P., Nagendra Prakash, S., Nagahanuman-tharao Suma, B.: Electrokinetic removal of heavy metals from soil. J. Electrochem. Sci. Eng. 5(1), 47–65 (2015). https://doi.org/10.5599/jese.2015.0055
96. Watson, M.C., Banuelos, G.S., O'Leary, J.W., Riley, J.J.: Trace element composition of Atriplex grown with saline drainage water. Agric. Ecosyst. Environ. **48**, 157–162 (1994)
97. Webb, K.L.: NaC1 effects on growth and transpiration in *Salicornia bigelovii*, a salt marsh halophyte. Plant Soil **24**, 261–268 (1966)
98. Wiszniewska et al.: Evaluation of the protective role of exogenous gro.pdf. (2018)
99. Wiszniewska et al.: Evaluation of the protective role of exogenous gro.pdf. (2018a)
100. Wiszniewska et al.: Evaluation of the protective role of exogenous gro.pdf. (2018b)
101. Wuana, R.A., Okieimen, F.E.: Heavy metals in contaminated soils: a review of sources, chemistry, risks and best available strategies for remediation. ISRN Ecol. **2011**, 1–20 (2011)
102. Yadav, N., Singh, S., Saini, O., Srivastava, S.: Environmental Nanotechnology, Monitoring & Management Technological advancement in the remediation of heavy metals employing engineered nanoparticles: a step towards cleaner water process. Environ. Nanotechnol., Monitor. Manag. **18**(November), 100757 (2022). https://doi.org/10.1016/j.enmm.2022.100757
103. Yasseen, B.T., Al-Thani, R.F.: Endophytes and halophytes to remediate industrial wastewater and saline soils: perspectives from Qatar. Plants **11**(11), 1497 (2022). https://doi.org/10.3390/plants11111497
104. Yensen, N.P. and Biel, K.Y.: Soil remediation via salt-conduction and the hypotheses of halosynthesis and photoprotection, tasks for vegetation science series 40. Ecophysiol. High Salinity Tolerant Plants 313–344 (2006)
105. Zeb, F., Archana, A. K. M., Gayathri, V., Revathy, K.: Remediation of heavy metals (Cr, Zn) using physical, chemical and biological methods : a novel approach. SN Appl. Sci. (2020). https://doi.org/10.1007/s42452-019-1918-x

106. Zhang, T., Liu, J.-M., Huang, X.-F., Xia, B., Su, C.-Y., Luo, G.-F., Xu, Y.-W., Wu, Y.-X., Mao, Z.-W., Qiu, R.-L.: Chelant extraction of heavy metals from contaminated soils using new selective EDTA derivatives. J. Hazard. Mater. **262**, 464–471 (2013)
107. ztürk, Z., Kose, D.A., Abdurrahman, A.S.A.N. and Ozkan, G.: Porous metal-organic Cu II complex of L-Arginine; synthesis, characterization, hydrogen storage properties and molecular simulation calculations. Hittite J. Sci. Eng. **1**(1), 1–5 (2014)

Investigation of Reacting Parameters to Improve Cationic Dye Adsorption onto Raw Pomegranate Peel

Rajae Ghibate, Meryem Ben Baaziz, Rachid Taouil, and Omar Senhaji

Abstract The current study investigated the ability of Moroccan pomegranate peel to reduce Methylene blue (MB) in aqueous media. Several reacting parameters were considered, including the dose of adsorbent, initial MB concentration, contact time, initial pH of the dye solution, ionic strength, as well as temperature. The experiments were conducted in a batch system to investigate MB adsorption. The results revealed a remarkably fast adsorption rate, with equilibrium being achieved within just 60 minutes. The MB adsorption is substantially impacted by the adsorbent dose, the initial pH of the solution, the presence of NaCl in the medium, and the initial dye concentration. The experimental results demonstrate that pomegranate peel exhibits a maximum adsorption capacity of 67.78 ± 0.70 mg/g when a dose of 2 g/L of pomegranate peel is used at a temperature of 25 °C. It was observed that as water temperature increased, MB adsorption decreased. Consequently, the findings indicate the exothermicity of the MB adsorption process. Moreover, the results strongly suggest that raw Moroccan pomegranate peel has promising potential as a biosorbent for treating dyestuff wastewater.

Keywords Methylene blue · Adsorption · Pomegranate peel · Reacting parameters

R. Ghibate (✉)
Laboratory of Physical Chemistry, Materials and Environment, Faculty of Sciences and Technologies, Moulay Ismail University, 52000 Errachidia, Morocco
e-mail: rajae.ghibate@gmail.com

M. B. Baaziz
Laboratory of Materials Engineering for the Environment and Natural Resources, Faculty of Sciences and Technologies, Moulay Ismail University, 52000 Errachidia, Morocco

R. Taouil
Laboratory of Mechanics, Energetics, Automation, and Sustainable Development, Faculty of Science and Technology, Moulay Ismail University, 52000 Errachidia, Morocco

O. Senhaji
Laboratory of Biomolecular and Macromolecular Chemistry, Moulay Ismail University of Meknès, 11201 Meknes, Morocco

J. Mabrouki and M. Azrour (eds.), *Integrated Solutions for Smart and Sustainable Environmental Conservation*, Studies in Systems, Decision and Control 527,
https://doi.org/10.1007/978-3-031-55787-3_12

1 Introduction

Methylene blue is one of the synthetic dyes frequently used in several spheres of our life, namely in medicine. Despite its low yield potential, methylene blue is utilized in textile industries to dye cotton, wool, and silk [1]. In addition, the release of MB into the environment leads to aesthetic concerns and dramatic changes in aquatic life [2]. That is in light of its visibility, low biodegradability, and bioaccumulation through the trophic chain. Indeed, the findings highlight the potential hazard to human health associated with acute exposure to MB. Such exposure can lead to various adverse effects, including tachycardia (rapid heart rate), jaundice (yellowing of the skin and eyes), quadriplegia (paralysis affecting all four limbs), tissue necrosis (cell death in body tissues), and cyanosis (bluish discoloration of the skin) [3]. It also leads to eye burns, which can permanently harm people and animals [4]. The literature has also shown additional negative consequences of MB, including hypertension, acute renal failure, and hemolytic anemia [5, 6]. Various conventional methods exist to remediate MB water pollution in this context. The frequently used include adsorption [7–13], chemical and advanced oxidation [14–17]. coagulation/flocculation [18–20], and membrane separation [21, 22]. Among these methods, adsorption is a promising and cheap alternative to wastewater treatment [23, 24]. Recently, different agricultural by-products were examined for dye removal, such as rice husk [3], eggshells [13], citrus peel [25], peanut hull [26], and banana peel [27]. This research seeks to enhance the pomegranate peel's effectiveness as an eco-friendly adsorbent against MB elimination from aqueous mediums. By studying and optimizing the adsorption process, the aim is to further harness the potential of pomegranate peel as a sustainable and efficient approach to treating dyestuff wastewater and mitigating the environmental impact of MB contamination. A preliminary study showed its potential use for the adsorption of MB. That prompted us to investigate the impact of more physicochemical parameters to improve the MB adsorption onto this locally abundant agricultural waste. In this respect, the current paper has focused on six reacting parameters likely to influence the MB adsorption onto pomegranate peel. It is about the dose of adsorbent, initial MB concentration, contact time, initial pH of the dye solution, ionic strength, and temperature.

2 Materials and Methods

2.1 Dyestuff Solution

Methylene blue ($C_{16}H_{18}ClN_3S$) was purchased. from Loba Chemie and employed as received, without requiring additional purification, for preparing the stock solution at a concentration of 1000 mg/L. By appropriately diluting, the MB working solutions were generated.

2.2 *Pomegranate Peel Preparation*

The pomegranate peel was stripped from the fruit, cleaned with distilled water, and sun-dried for about three weeks until it turned dark. The dried fragments were crushed, and the powder was repeatedly rinsed with distilled water to produce a colorless washing solution. After 48 hours of drying at 60 °C, the pomegranate peel powder was placed in a desiccator to await usage.

2.2.1 Experiments

The batch system was considered for studying MB adsorption according to pomegranate peel dose, contact time, initial dye solution pH, ionic strength, initial MB concentration, and temperature. Each experiment was conducted by adding a specified mass of prepared pomegranate peel powder to 50 mL of the MB solution in a 100 mL conical flask. The contents in the flask were shaken at 300 rpm by an electromagnetic stirrer. During the course of the study, several parameters were considered to understand the pomegranate peel adsorption behavior. It is about the dose of pomegranate peel powder (0.25–6 g/L), the adsorption time (0–300 min), the dye solution's initial pH (3–11), the concentration of NaCl (0–0.1 mol/L), MB initial concentration (50–600 mg/L), plus the temperature (25–55 °C).

The pH adjustment was assured by HNO_3 or NaOH solutions. After adsorption, the suspensions were subjected to centrifugation at 3800 rpm for about five minutes. The MB residual concentration in the supernatant was determined at 665 nm using double-beam UV–Vis spectrophotometers, then exploited in the following equation to get the adsorption capacity:

$$q_t = \frac{(C_0 - C_t)}{m} \cdot V \tag{1}$$

where C_0 and C_t (mg/g) represent respectively the MB concentration in the solution before and after adsorption at the desired time (mg/L); m (g) refers to pomegranate peel powder mass (g), and V (L) indicates the volume of dye solution (L).

3 Results and Discussion

3.1 *Influence of Adsorbent Dose*

Figure 1 depicts the impact of pomegranate peel dose on the MB adsorption from aqueous solutions. Adsorbent doses ranged from 0.25 to 6 g/L in experiments under initial conditions of 100 mg/L of MB, unadjusted pH, at 25 °C for 120 min. The findings reveal that when the pomegranate peel dose rises from 0.25 to 6 g/L, the

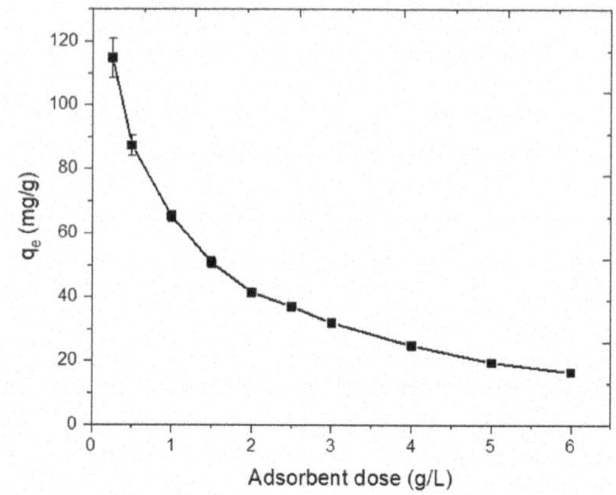

Fig. 1 Adsorbent dose effect on MB adsorption onto pomegranate peel

amount of MB adsorbed drops significantly. That can be assigned to the existence of unsaturated adsorption sites during the process of adsorption [28]. The abundance of the adsorbent in the aqueous medium may cause particle aggregation and then a restriction in the molecules of MB adsorbed. Furthermore, 2 g/L was selected as the optimum dose to investigate other parameters' effects.

3.2 *Influence of Contact Time*

The reaction time effect on the uptake of MB at the surface of pomegranate peel was evaluated at 100 mg/L of MB concentration, unadjusted pH, 2 g/L of adsorbent dose, and 25 °C. The obtained results are represented in Fig. 2. It comes out clearly that the adsorption process consists of two steps. The rate of adsorbed MB rose quickly in the first one, reaching 42.71 ± 0.22 mg/g after just 10 min. After that, the amount of MB adsorbed rose progressively until reaching a maximum at 60 min. Hereafter, no significant changes were noticed. This behavior can be assigned to the fact that there were initially several vacant sites on the surface of the pomegranate peel and that these sites gradually filled up with MB molecules until they were saturated. The same behavior was seen in earlier research [6, 29, 30]. The reaction time will be fixed at 120 min for further adsorption experiments.

3.3 *Influence of Initial pH*

pH effect on MB adsorption was evaluated at initial pH ranges of 3–11 under the initial conditions of 100 mg/L of initial MB concentration, 2 g/L of pomegranate peel, 25 °C, and 120 min of reaction time. pH adjustment was performed with either HNO_3 or NaOH solutions. Figure 3 illustrates that the amount of MB adsorbed improves

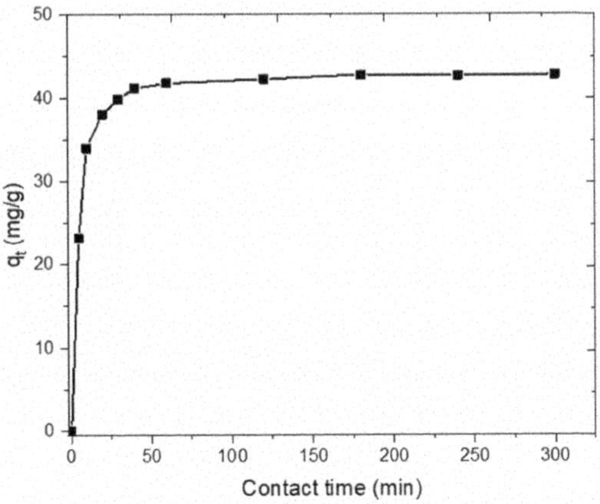

Fig. 2 Contact time effect on MB adsorption onto pomegranate peel

further with rising pH, peaking around pH 11. It is possible that dye absorption can be low at low pH values due to the competition for absorption sites between H^+ ions and cationic dye molecules. Then, increasing pH decreased the competitive effect of H^+, and dye cations took over the free-binding sites [31]. Moreover, at pH values exceeding the pH value at the point of zero charge ($pH_{pzc} = 4.95$), the pomegranate peel surface is negatively charged, leading to an electrostatic attraction between the adsorption sites and the cationic dye, which enhances MB uptake. While at a pH lower than pH_{pzc}, the surface of the pomegranate peel become positively charged, which reduces the absorption of MB due to electrostatic repulsion.

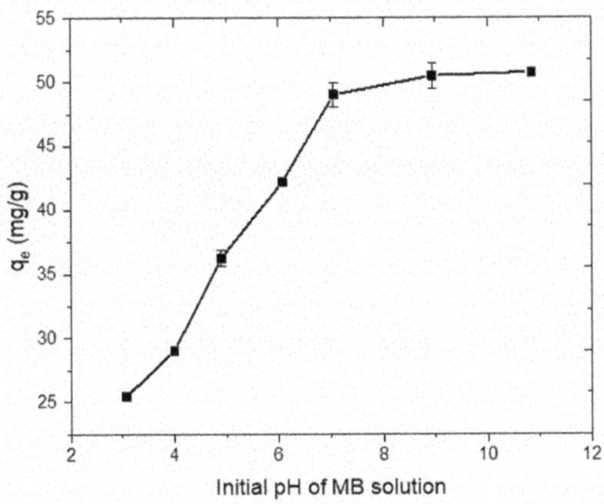

Fig. 3 Initial pH effect on MB adsorption onto pomegranate peel

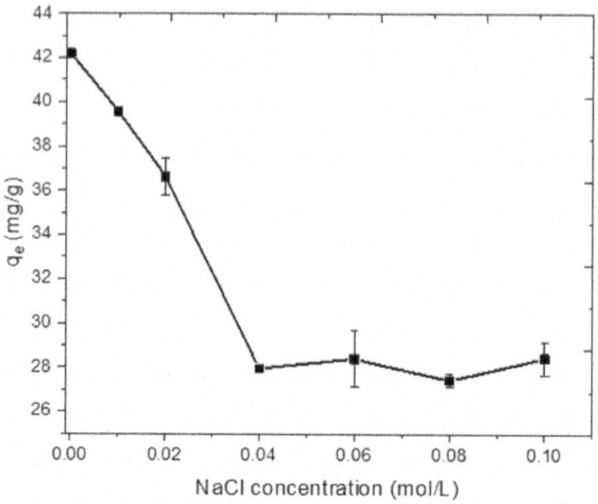

Fig. 4 NaCl concentration effect on MB adsorption onto pomegranate peel

3.4 Influence of Ionic Strength

Furthermore, the ionic strength impact on MB adsorption was investigated. The experiments were conducted at 100 mg/L of MB and 2 g/L of pomegranate peel at an unadjusted pH for 120 min at varied NaCl concentrations from 0 to 0.1 mol/L. Figure 4 illustrates how adding NaCl to the dye solution reduced its adsorption. The MB adsorbed onto pomegranate peel decreased from 42.20 $\pm$ 0.18 mg/g to 28.43 $\pm$ 0.76 mg/g by enriching the dye solution with NaCl content from 0 to 0.10 mol/L. Indeed, in the presence of NaCl, two factors might account for the decrease in MB adsorbed: first, the decrease of electrostatic attraction between the MB and the binding sites, and second, the competitive impact of Na^+ on binding sites. When NaCl is present in large quantities (0.10 mol/L) in the dye solution, Na^+ tends to saturate the binding sites in the adsorbent, which restricts MB access to the adsorbent surface. Furthermore, the electrical diffused double layer grows or expands when NaCl is present. Such expansion may prevent the adsorbent and cationic dye from getting closer to each other, which reduces the electrostatic attraction [32].

3.5 Influence of Initial Concentration of Dye

An experimental study was conducted over a 50–600 mg/L concentration range to deduce the consequences of the initial dye concentration on MB adsorption onto pomegranate peel. As illustrated in Fig. 5, MB adsorption rises as the concentration of dye increases up to 500 mg/L. Beyond this concentration, the amount of MB adsorbed becomes relatively constant, presenting a saturation trend owing to a finite

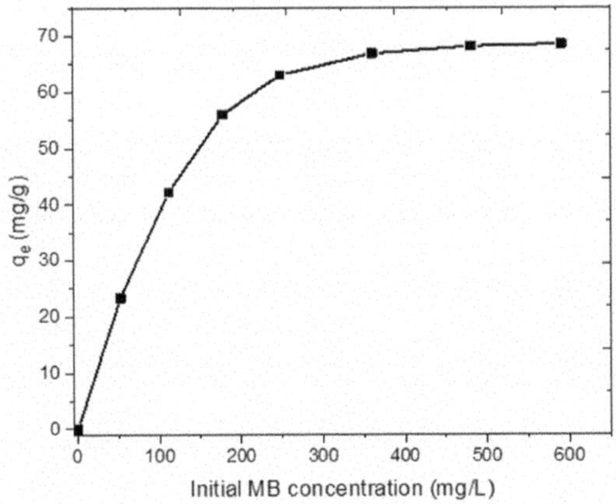

Fig. 5 Initial dye concentration effect on MB adsorption onto pomegranate peel

number of surface binding sites at higher MB concentrations. As a result, 67.78 ± 0.70 mg/g was the highest adsorption capacity. The driving force produced by the concentration gradient may be the cause of the increase in adsorption capacity that occurs as MB concentration rises. That heightens dye molecules' diffusion across the adsorbent surface. Other researchers employing different adsorbents have noticed a similar trend [33, 34].

3.6 *Influence of Temperature*

The temperature impact on methylene blue fixation at the surface of pomegranate peel was evaluated between 25 and 55 °C, letting the reaction proceed for 120 min. The initial experimental conditions consisted of using a 100 mg/L of MB concentration, 2 g/L of pomegranate peel dose, keeping the pH of MB solution unadjusted. Figure 6 depicts that the adsorbed amount of MB onto pomegranate peel dropped from 42.20 ± 0.18 to 38.44 ± 0.01 mg/g as the temperature grew from 25 to 55 °C. This finding points to the exothermicity of the MB adsorption process.

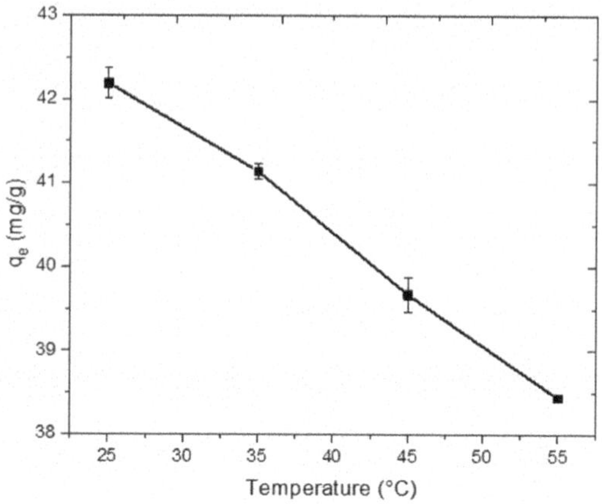

Fig. 6 Temperature effect on MB adsorption onto pomegranate peel

4 Conclusion

By carefully controlling various parameters during batch experiments, the methylene blue adsorption onto pomegranate peel was significantly improved. Several factors were identified as influential: adsorbent dose, contact time, initial pH of the solution, ionic strength, initial dye concentration, and temperature. The adsorption process exhibited rapid kinetics, reaching equilibrium in 60 min. Notably, the adsorption capacity of pomegranate peel for MB was greatly enhanced at higher pH levels, favoring alkaline conditions. Additionally, a decrease in temperature resulted in higher MB adsorption, suggesting the exothermicity of the adsorption process.

Through optimization, it was determined that an optimal adsorbent dose of 2 g/L yielded a considerable adsorption capacity for pomegranate peel at 25 °C, measuring 67.78 ± 0.70 mg/g. These findings highlight the potential of pomegranate peel as an effective and green approach to removing MB dye from water systems.

References

1. Muhammad, A.S., Abdurrahman, M.A.: Adsorption of methylene blue onto modified agricultural waste. Moroccan J. Chem. **8**(2), 412–427 (2020)
2. Elmontassir, H., Elfalaki, K., Karhat, Y., Afdali, M.: Adsorption of a dye and a real rejection of textile on sludge from drinking water treatement. Moroccan J. Chem. **7**(3), 493–505 (2019)
3. Vadivelan, V., Kumar, K.V.: Equilibrium, kinetics, mechanism, and process design for the sorption of methylene blue onto rice husk. J. Colloid Interface Sci. **286**(1), 90–100 (2005)
4. Sen, T.K., Afroze, S., Ang, H.M.: Equilibrium, kinetics and mechanism of removal of methylene blue from aqueous solution by adsorption onto pine cone biomass of Pinus radiata. Water Air Soil Pollut. **218**(1), 499–515 (2011)
5. Shannon, M.W., Borron, S.W., Burns, M.J., Haddad, L.M., Winchester, J.F. (eds.): Haddad and Winchester's Clinical Management of Poisoning and Drug Overdose, 4th edn. Elsevier, United States (2007)

6. Cheng, Z., Yang, R., Zhu, X.: Adsorption behaviors of the methylene blue dye onto modified sepiolite from its aqueous solutions. Desalin. Water Treat. **57**(52), 25207–25215 (2016)
7. Regti, A., Tourit, A., Laamari, R., Haddad, M.E.: Enhancement removal of Basic Blue 9 dye from wastewaters using natural Moroccan wood as efficient adsorbent. Moroccan J. Chem. **2**(1), 22–32 (2014)
8. Jodeh, S., Amarah, J., Radi, S., Hamed, O., Warad, I., Salghi, R., Chetouani, A., Samhan, S., Alkowni, R.: Removal of methylene blue from industrial wastewater in Palestine using polysiloxane surface modified with bipyrazolic tripodal receptor. Moroccan J. Chem. **4**(1), 140–156 (2016)
9. Badraoui, A.E., Miyah, Y., Nahali, L., Zerrouq, F., Khazzan, B.E.: Fast adsorption for removal of methylene blue from aqueous solutions using of local clay. Moroccan J. Chem. **7**(3), 416–423 (2019)
10. Badri, N., Chhiti, Y., Bentiss, F., Bensitel, M.: Removal of cationic dye by high surface activated carbon prepared from biomass (date pits) by carbonization and activation processes. Moroccan J. Chem. **6**(4), 767–780 (2018)
11. Otheman, A., Hakima, N., Eddine, M.S.E.: Modeling of adsorption isotherms of methylene blue onto natural illitic clay: nonlinear regression analysis. Moroccan J. Chem. **3**(3), 582–593 (2015)
12. Kesraoui, A., Seffen, M., Brouers, F.: Non linear modelisation of dyes removal from aqueous solution by using sorption onto Luffa cylindrica fibers. Moroccan J. Chem. **5**(4), 659–676 (2017)
13. Akazdam, S., Yassine, W., Chafi, M., Gourich, B.: A comparative study for removal of methylene blue by NaOH treated and raw eggshells. Moroccan J. Chem. **7**(2), 300–313 (2019)
14. Dutta, K., Mukhopadhyay, S., Bhattacharjee, S., Chaudhuri, B.: Chemical oxidation of methylene blue using a Fenton-like reaction. J. Hazard. Mater. **84**(1), 57–71 (2001)
15. Al Jibouri, A.K.H., Wu, J., Upreti, S.R.: Continuous ozonation of methylene blue in water. J. Water Process Eng. **8**, 142–150 (2015)
16. Liu, Y., Jin, W., Zhao, Y., Zhang, G., Zhang, W.: Enhanced catalytic degradation of methylene blue by α-Fe_2O_3/graphene oxide via heterogeneous photo-Fenton reactions. Appl. Catal. B **206**, 642–652 (2017)
17. Hama Aziz, K.H., Mahyar, A., Miessner, H., Mueller, S., Kalass, D., Moeller, D., Khorshid, I., Rashid, M.A.M.: Application of a planar falling film reactor for decomposition and mineralization of methylene blue in the aqueous media via ozonation, Fenton, photocatalysis and non-thermal plasma: a comparative study. Process Saf. Environ. Prot. **113**, 319–329 (2018)
18. Lau, Y.-Y., Wong, Y.-S., Teng, T.-T., Morad, N., Rafatullah, M., Ong, S.-A.: Degradation of cationic and anionic dyes in coagulation–flocculation process using bi-functionalized silica hybrid with aluminum-ferric as auxiliary agent. RSC Adv. **5**(43), 34206–34215 (2015)
19. Othmani, A., Kesraoui, A., Seffen, M.: The alternating and direct current effect on the elimination of cationic and anionic dye from aqueous solutions by electrocoagulation and coagulation flocculation. Euro-Mediterr. J. Environ. Integr. **2**(1), 1–12 (2017)
20. Wei, Q., Mcyotto, F.O., Chow, C.W.K., Nadeem1, Z., Li, Z., Liu, J., Desk, S.: Eco-friendly decolorization of cationic dyes by coagulation using natural coagulant Bentonite and biodegradable flocculant Sodium Alginate. J. Earth Sci. Environ. Stud. **5**(2), 51–60 (2020)
21. Li, Q., Li, Y., Ma, X., Du, Q., Sui, K., Wang, D., Wang, C., Li, H., Xia, Y.: Filtration and adsorption properties of porous calcium alginate membrane for methylene blue removal from water. Chem. Eng. J. **316**, 623–630 (2017)
22. Cheng, J., Zhan, C., Wu, J., Cui, Z., Si, J., Wang, Q., Peng, X., Turng, L.-S.: Highly efficient removal of methylene blue dye from an aqueous solution using cellulose acetate nanofibrous membranes modified by polydopamine. ACS Omega **5**(10), 5389–5400 (2020)
23. Ghibate, R., Sabry, F., Senhaji, O., Taouil, R., Touzani, M.: State of the art of technologies for Zn^{2+} ions removal from industrial effluents with adsorption: examination of process parameters (Part I). Int. J. Innov. Res. Sci. Technol. **2**(7), 39–48 (2015)
24. Rozumová, L., Životský, O., Seidlerová, J., Motyka, O., Šafařík, I., Šafaříková, M.: Magnetically modified peanut husks as an effective sorbent of heavy metals. J. Environ. Chem. Eng. **4**(1), 549–555 (2016)

25. Rani, S., Chaudhary, S.: Adsorption of methylene blue and crystal violet dye from waste water using *Citrus limetta* peel as an adsorbent. Mater. Today: Proc. **60**, 336–344 (2022)
26. Song, J., Zou, W., Bian, Y., Su, F., Han, R.: Adsorption characteristics of methylene blue by peanut husk in batch and column modes. Desalination **265**(1), 119–125 (2011)
27. Hashem, A.H., Saied, E., Hasanin, M.S.: Green and ecofriendly bio-removal of methylene blue dye from aqueous solution using biologically activated banana peel waste. Sustain. Chem. Pharm. **18**, 100333 (2020)
28. Zhu, H.-X., Cao, X.-J., He, Y.-C., Kong, Q.-P., He, H., Wang, J.: Removal of Cu^{2+} from aqueous solutions by the novel modified bagasse pulp cellulose: kinetics, isotherm and mechanism. Carbohyd. Polym. **129**, 115–126 (2015)
29. Jia, Z., Li, Z., Ni, T., Li, S.: Adsorption of low-cost absorption materials based on biomass (*Cortaderia selloana* flower spikes) for dye removal: kinetics, isotherms and thermodynamic studies. J. Mol. Liq. **229**, 285–292 (2017)
30. Rangabhashiyam, S., Lata, S., Balasubramanian, P.: Biosorption characteristics of methylene blue and malachite green from simulated wastewater onto *Carica papaya* wood biosorbent. Surf. Interfaces **10**, 197–215 (2018)
31. Banat, F., Al-Asheh, S., Al-Makhadmeh, L.: Evaluation of the use of raw and activated date pits as potential adsorbents for dye containing waters. Process Biochem. **39**(2), 193–202 (2003)
32. Ngah, W.S.W., Hanafiah, M.A.K.M.: Biosorption of copper ions from dilute aqueous solutions on base treatedrubber (*Hevea brasiliensis*) leaves powder: kinetics, isotherm, and biosorption mechanisms. J. Environ. Sci. **20**(10), 1168–1176 (2008)
33. Dawood, S., Sen, T.K.: Removal of anionic dye Congo red from aqueous solution by raw pine and acid-treated pine cone powder as adsorbent: equilibrium, thermodynamic, kinetics, mechanism and process design. Water Res. **46**(6), 1933–1946 (2012)
34. Boumehdi Toumi, L., Hamdi, L., Salem, Z., Allia, K.: Batch adsorption of methylene blue from aqueous solutions by untreated Alfa grass. Desalin. Water Treat. **53**(3), 806–817 (2015)

Performance Evaluation of Irrigation Projects: The Power of Remote Sensing Tools

Ahmed Elshaikh, Jamal Mabrouki, and Mourade Azrour

Abstract Irrigation performance concerns with knowing the extent to which an irrigation system achieved its established objectives. In this regard, different methodologies and tools have been developed to assess and evaluate irrigation projects performance. Remote Sensing (RS) has become a major tool for assessing and evaluating irrigation systems performance. Remote sensing management tools provide a system-wide and spatially distributed information for the evaluation process. The main concept is based on derived satellite images data that can be merged with maps of the irrigation delivery systems to find real time performance. This paper aims to identify the advantages and disadvantages of RS as a tool with high potentials in performance evaluation of irrigation projects. In addition, the paper provides new insights and recommendations to improve the performance of irrigation systems in terms of water saving and increasing crops productivity.

Keywords Performance evaluation · Irrigation · Remote sensing

1 Introduction

The world's population is expanding rapidly and there will be a rise in the need for food. More effective and efficient use of land and water resources is necessary to accomplish this. Irrigation projects typically operate much below their capacity, regardless of their potential for food production, which results in a remarkable decline

A. Elshaikh (✉)
Water Research Center, University of Khartoum, Khartoum, Sudan
e-mail: ahmedhayaty@live.com

J. Mabrouki
Laboratory of Spectroscopy, Molecular Modeling, Materials, Nanomaterials, Water and Environment, Faculty of Science, CERNE2D, Mohammed V University in Rabat, Rabat, Morocco

M. Azrour
Engineering Science and Technology Laboratory, IDMS Team, Faculty of Sciences and Techniques, Moulay Ismail University of Meknes, Errachidia, Morocco

J. Mabrouki and M. Azrour (eds.), *Integrated Solutions for Smart and Sustainable Environmental Conservation*, Studies in Systems, Decision and Control 527,
https://doi.org/10.1007/978-3-031-55787-3_13

in performance. The lack of anticipated advantages, poor health and environmental effects, and ineffective resource management are all factors contributing to this predicament [9, 10]. The complexity of the problem has led to an increase in initiatives meant to boost the effectiveness of irrigation projects. Large investments in irrigation projects have sparked concerns about the effectiveness of irrigation and the fairness of the advantages that it provides. It is generally concluded that the economic performance of irrigation has been considerably lower than expected by its proponents, even though there are many cases of positive and even high internal rates of return [12]. In addition, there has been concern that smallholders have been bypassed by many irrigation developments, as well as fears that many of the benefits have been captured by larger landowners [1]. Performance evaluation tools could assist in validating the most appropriate project lessons learned and creating benchmarks to enhance planning, implementation, and administration of similar projects in order to increase the productivity of these projects. The goal of the performance evaluation is to ensure that resources are used efficiently and effectively by giving the scheme management at all levels key inputs. Additionally, it will help in collecting information so that remedial measures can be implemented to maximize the advantages of the irrigation project [10]. The procedure for conducting performance reviews is not linear because numerous tasks must be completed simultaneously and in a systematic way must be coordinated within the constraints of the resources that happen to be available. A lot of work has been done to evaluate the results of such interventions or to improve understanding of performance, in order to improve this process even more [3, 10, 18, 19]. The Analysis Hierarchy Process (AHP), Fuzzy set theory, direct measurements for indicators, and remote sensing are some of the methodologies from the literature used to analyze irrigation system performance [11]. These techniques are capable of helping in validating the pertinent project lessons discovered and creating benchmarks to enhance project planning, execution, and management [11]. This chapter's goal is to outline the benefits and drawbacks of using RS as a technique to assess the performance of irrigation projects. The chapter also offers new perspectives and recommendations to improve irrigation system efficiency in terms of both water conservation and crops productivity.

2 Remote Sensing

According to Thenkabail et al. [20], RS could be defined as "the measurement of object properties on the earth's surface using data acquired from aircraft and satellites". As a result, it is an attempt to measure something remotely rather than on-site. We must rely on transferring signals of some kind, such as optical, acoustic, or microwave waves, as we are not in direct contact with the item of interest (Table 1).

Data from remote sensing can be in the form of profiles along flight paths or individual point measurements. Many remote-sensing systems have been created, giving a wide variety of geographical, spectral, and temporal parameters, to fulfill the needs of various data users (Table 2).

Table 1 The main spectral bands used for remote sensing of the earth [17]

Name	Surface property of interest	Radiation source	Wavelength range
Visible (V)	Reflectance	Solar	0.4–0.7 μm
Near infrared (NIR)	Reflectance	Solar	0.7–1.1 μm
Short wave infrared (SWIR)	Reflectance	Solar	1.1–1.35 μm 1.4–1.8 μm 2–2.5 μm
Midwives infrared (MWIR)	Reflectance, temperature	Solar, Thermal	3–4 μm 4.5–5 μm
Thermal or long wave infrared (TIR or LWIR)	Temperature	Thermal	8–9.5 μm 10–14 μm
Microwave, radar	Temperature (passive), roughness (active)	Thermal (passive), artificial (active)	1 mm–1 m

Table 2 Primary geophysical variables and measurable parameters [17]

General description	Specific descriptions
Land/cloud boundaries	Cloud and vegetation Vegetation chlorophyll
Land/cloud properties	Soil, vegetation differences Green vegetation Leaf-canopy properties Snow-cloud differences
Ocean color	Chlorophyll observations-flourescence Sediments, atmosphere Aerosol—atmosphere properties
Atmosphere/clouds	Cloud-atmosphere properties
Thermal properties	Sea surface temperatures Forest fires-volcanoes Cloud—surface temperature
	Troposphere temp—cloud fraction
Atmosphere/clouds	Cirrus clouds
Thermal properties	Mid & upper—troposphere humidity Surface temperature
	Total ozone
	Cloud—surface temperature
	Cloud height and fraction

3 Remote Sensing for Irrigation Performance

The performance of irrigation systems can be easily gauged and evaluated using remote sensing (RS). To determine real-time performance, RS generated satellite photos that can be combined with data from irrigation delivery systems [7]. As well as offering the chance to examine crop growing characteristics at scales ranging from individual farms to large-scale initiatives, remote sensing could be utilized to improve diagnosis in data-scarce scenarios [8, 22].

3.1 The Technical Characteristics of the Remote Sensing

Large scale and geographically dispersed information are well represented through RS management tools. The Surface Energy Balance Algorithm for Land (SEBAL), developed to solve the energy balance components pixel-by-pixel, is the most popular and effective tool of the remote sensing evapotranspiration algorithm [2]. Physically, it is based on the utilization of net radiation fraction and soil heat flow [15]. To ascertain the constitutive characteristics of SEBAL, scanned spectral radiances in the visible, near-infrared, and thermal infrared region are required (Fig. 1).

Evapotranspiration is computed using the information provided by dividing the latent heat flux by the daily average net radiation. On a pixel-by-pixel basis, the latent heat flux is calculated directly from the surface energy balance at satellite overpass as shown below:

$$\lambda E = R_n - (G_0 + H) \tag{1}$$

where:

R_n = the net radiation.

G_0 = the soil heat flux.

H = the sensible heat flux.

λE = the latent heat flux.

On the other hand, the density of green cover on a piece of land is typically assessed using the Normalized Difference Vegetation Index (NDVI). According to Kogan [13] and Wan et al. [21] unhealthy or sparse vegetation reflects more visible light and less near-infrared light. As shown below, the Normalized Difference Vegetation Index (NDVI) is calculated as:

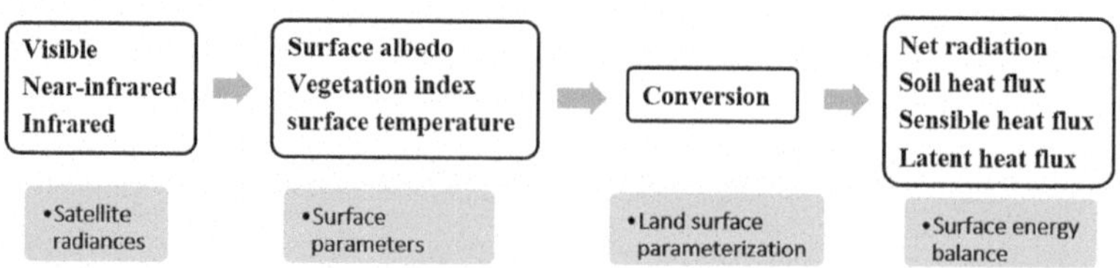

Fig. 1 The surface energy balance algorithm [15]

$$NDVI = \frac{NRI - R}{NRI + R} \tag{2}$$

where:

NIR = the Near Infra-Red radiation reflected by vegetation.

R = visible radiation reflected by vegetation.

NDVI = the Normalized Difference Vegetation Index.

3.2 *Application of RS (Case Studies)*

RS tools were used in several research to assess irrigation performance. Roerink et al. [15] studied how to calculate water efficiency for the Rio Tunuyan irrigation system in Argentina as one of the initial investigations that employed remote sensing techniques. Using Advanced Very High Resolution Radiometer (AVHRR) data from the National Oceanic and Atmospheric Administration (NOAA), Bastiaanssen and Ali [6] calculated crop production in the Indus Basin, Pakistan. Similar methods were used by Hamid et al. [12] and Ahmad et al. [2], to diagnose irrigation performance of the Rahad Scheme in Sudan and the Rachna Doab project in Pakistan, respectively, using NOAA-AVHRR satellite pictures.

Using remote sensing methods, Zwart and Leclert [23] assessed irrigation performance of the Office du Niger's project in Mali. Using satellite data, Mamad [14] evaluated the productivity of the land and water at several spatial scales within the Gezira scheme, Sudan. Bandara [4] did another study utilizing satellite imagery to evaluate the water productivity for three major irrigation systems in Sri Lanka. Using satellite data and remote sensing, Bashir et al. [5] and Santos et al. [16], computed crop water requirements and evaluated irrigation efficiency from the spatial seasonal actual evapotranspiration (ET).

The above studies used several RS methods to examine the efficient operation of irrigation systems from different perspectives.

4 WaPOR Portal

The Food and Agriculture Organization (FAO) developed WaPOR, an online platform for tracking water productivity using openly accessible remotely sensed data. The tool seeks to track and report on water production in agriculture. The water productivity database and its thousands of underlying map layers are accessible to everyone.

Through a specific FAO WaPOR, users can immediately access the portal's services, which are also published and documented on the FAO website.

It enables direct data inquiries, time series analysis, geographical statistics, and data downloads of essential variables related to evaluations of water and land productivity. Below are some examples of these data.

4.1 Climate

The climate data include precipitation and reference evapotranspiration. These data are available at different temporal resolutions: daily, decadal, monthly and annual, (The precipitation has resolution of 5 km and reference evapotranspiration of 20 km) (Fig. 2).

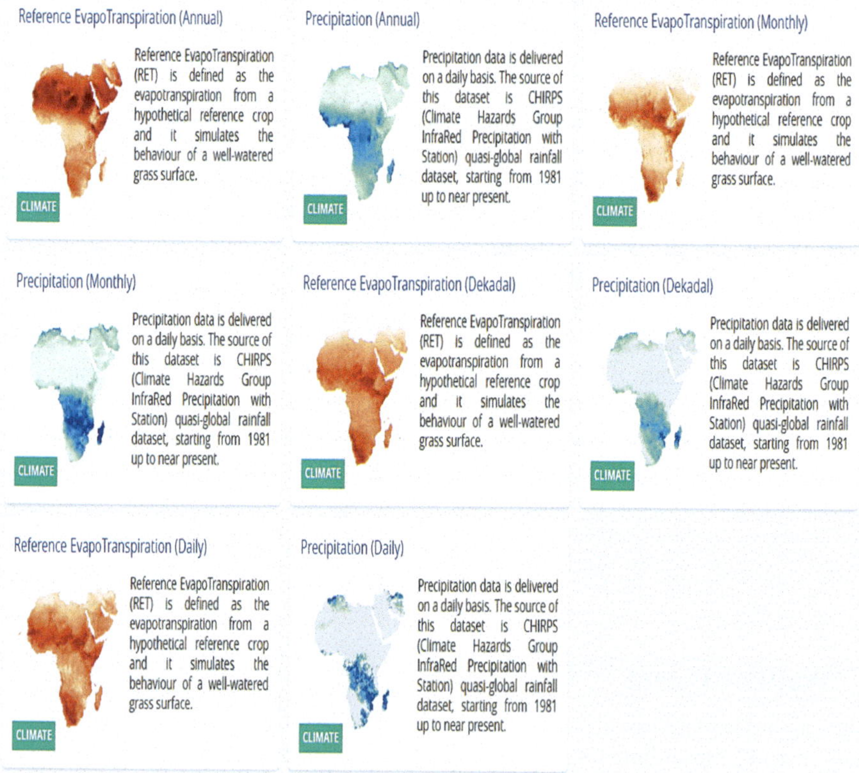

Fig. 2 RS climate data include precipitation and reference evapotranspiration (*Source* WAPOR)

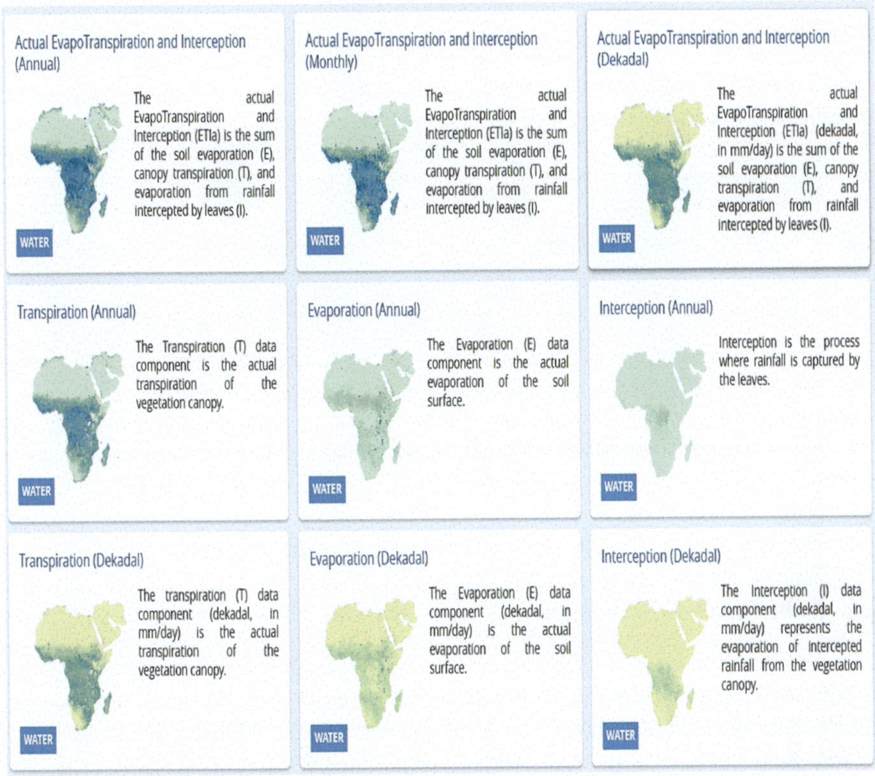

Fig. 3 RS water data include evaporation, transpiration, and interception (*Source* WAPOR)

4.2 Water

Water data include Evaporation, transpiration, Interception and sum of these three. These data are available at all three levels with decadal and annual temporal resolution. In addition, monthly actual evapotranspiration and interception are also available (Fig. 3).

4.3 Land

Land data include "net primary production", "total biomass production", "land cover classification and phenology". Decadal net primary production is available for the three levels. Total biomass production is available annually at level one and seasonally at level two and three.

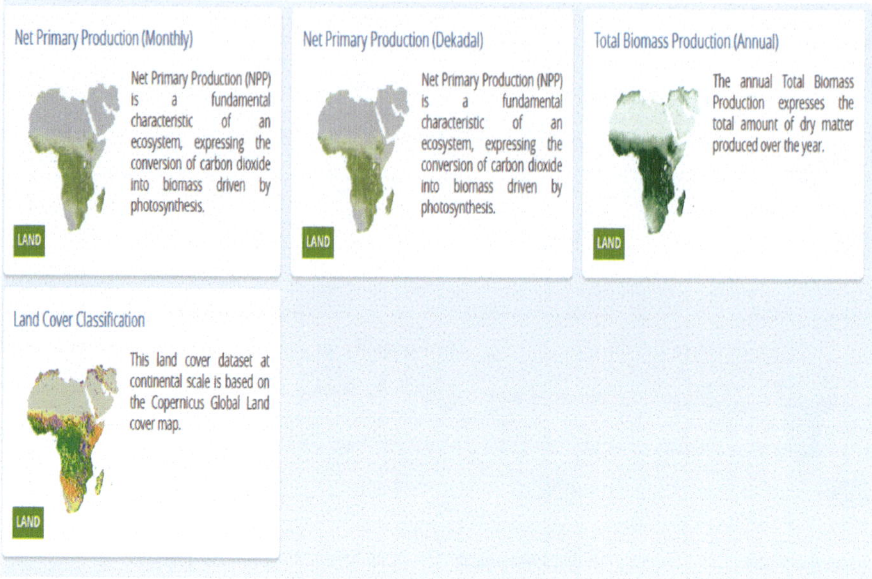

Fig. 4 RS land data from WAPOR platform (*Source* WAPOR)

Similarly, phenology is available seasonally at level two and three. Land cover classification is available annually at level one and two. In addition, for some areas Decadal land cover classification with information on the crop types is available (Fig. 4).

4.4 The Water Productivity

The Water productivity data is computed from biomass production and water data components. These are available annually at level one and seasonally at level two and three (Fig. 5).

5 Remote Sensing's Advantages and Drawbacks

Remote sensing has enormous potential since it makes it simple to gather data at different scales and resolutions. It is possible to analyze and understand a single image obtained through remote sensing for usage in a variety of applications and objectives. Additionally, remotely sensed data can be quickly processed and analyzed on a computer and used for a variety of tasks.

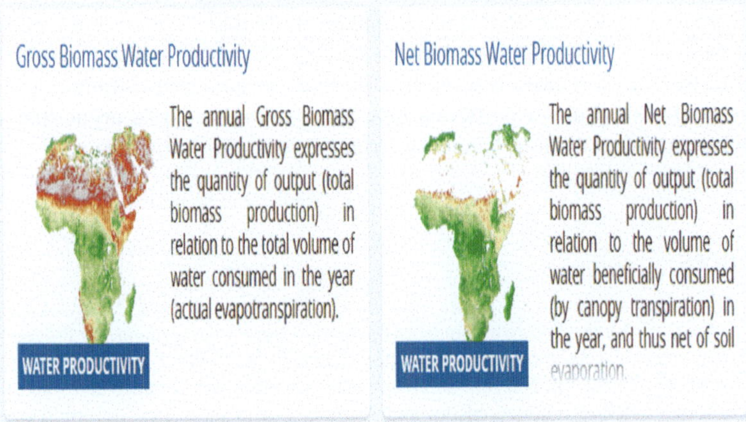

Fig. 5 RS water productivity data is computed from biomass production and water data components (*Source* WAPOR)

Additionally, repetitive coverage is affordable, which is helpful when gathering information on dynamic subjects like water and agricultural field data. Additionally, it makes it possible to collect data for locations that might be challenging or impossible to sample due to logistical or financial limitations.

On the other hand, there is no technology without flaws. When measuring or evaluating smaller regions, remote sensing methodology is more expensive but produces data that are more exact and detailed. The technology is expensive and necessitates specialized training to examine the images. For accurate readings and to prevent errors, remote sensing data require equipment that is regularly calibrated. Despite these difficulties, remote sensing has more advantages than obstacles, and scientific research and technological advancement are predicted to make it better.

6 Conclusion

Through remote sensing, water managers can determine where scarce water resources are wasted and where water productivity might be increased in a specific area. Concerns about irrigation performance include determining the degree to which an irrigation system met its predetermined goals. In this regard, several approaches and instruments have been created to gauge the effectiveness of irrigation schemes. The performance of irrigation systems can now be assessed and evaluated using remote sensing (RS). System-wide and spatially distributed information is made available for the evaluation process by remote sensing technologies. The basic concept is based on satellite image data that has been deduced and that may be combined with maps of irrigation delivery systems to determine real-time performance.

Crop classification, area of crops estimation, and yield evaluation are all crucial arenas in which remote sensing is extremely important. Additionally, it enables scientists and farmers to forecast crop production over an area and estimate how much crop may be gathered under particular circumstances. One could argue that spatial data will be crucial to the assessment of irrigation systems in the future.

References

1. Abernethy, C.L.: Methodology for studies of irrigation water management. In: Proceedings of the African Regional Symposium on Smallholder Irrigation (1984)
2. Ahmad, M.-D., Turral, H., Nazeer, A.: Diagnosing irrigation performance and water productivity through satellite remote sensing and secondary data in a large irrigation system of Pakistan. Agric. Water Manag. **96**, 551–564 (2009)
3. Alcon, F., Garc, Bastida, P.A., Soto-Garc, M., Mart, inez-Alvarez, V., Martin-Gorriz, B., Baille, A.: Explaining the performance of irrigation communities in a water-scarce region. Irrig. Sci. 1–11 (2017)
4. Bandara, K.: Monitoring irrigation performance in Sri Lanka with high-frequency satellite measurements during the dry season. Agric. water Manag. **58**, 159–170 (2003)
5. Bashir, M.A., Hata, T., Tanakamaru, H., Abdelhadi, A.W., Tada, A.: Satellite-based energy balance model to estimate seasonal evapotranspiration for irrigated sorghum: a case study from the Gezira scheme, Sudan. Hydrol. Earth Syst. Sci. **12**, 1129–1139 (2008)
6. Bastiaanssen, W.G.M., Ali, S.: A new crop yield forecasting model based on satellite measurements applied across the Indus Basin, Pakistan. Agric. Ecosyst. Environ. **94**, 321–340 (2003)
7. Bastiaanssen, W.G.M., Bos, M.G.: Irrigation performance indicators based on remotely sensed data: a review of literature. Irrig. Drain. Syst. **13**, 291–311 (1999)
8. Bastiaanssen, W.G.M., Molden, D.J., Makin, I.W.: Remote sensing for irrigated agriculture: examples from research and possible applications. Agric. Water Manag. **46**, 137–155 (2000)
9. Biswas, A.K.: Monitoring and evaluation of irrigation projects. J. Irrig. Drain. Eng. **116**, 227–242 (1990)
10. Bos, M.G., Burton, M.A.S., Molden, D.J.: Irrigation and Drainage Performance Assessment: Practical Guidelines. CABI (2005)
11. Elshaikh, A.E., Jiao, X., Yang, S.: Performance evaluation of irrigation projects: theories, methods, and techniques. Agric. Water Manag. **203**, 87–96 (2018). https://doi.org/10.1016/j.agwat.2018.02.034
12. Hamid, S.H., Mohamed, A.A., Mohamed, Y.A.: Towards a performance-oriented management for large-scale irrigation systems: case study, Rahad scheme, Sudan. Irrig. Drain. **60**, 20–34 (2011)
13. Kogan, F.N.: Global drought watch from space. Bull. Am. Meteorol. Soc. **78**, 621–636 (1997). https://doi.org/10.1175/1520-0477(1997)078%3c0621:GDWFS%3e2.0.CO;2
14. Mamad, N.: Irrigation Performance of Gezira Scheme, Sudan: Assessment of Land and Water Productivity Using Satellite Data. Unesco-IHE (2010)
15. Roerink, G.J., Bastiaanssen, W.G.M., Chambouleyron, J., Menenti, M.: Relating crop water consumption to irrigation water supply by remote sensing. Water Resour. Manag. **11**, 445–465 (1997)
16. Santos, C., Lorite, I.J., Tasumi, M., Allen, R.G., Fereres, E.: Integrating satellite-based evapotranspiration with simulation models for irrigation management at the scheme level. Irrig. Sci. **26**, 277–288 (2008)
17. Schowengerdt, R.A.: Remote Sensing: Models and Methods for Image Processing. Elsevier (2006)

18. Small, L.E., Svendsen, M.: A Framework for Assessing Irrigation Performance. Washington, D.C. (1992)
19. Sun, H., Wang, S., Hao, X.: An improved analytic hierarchy process method for the evaluation of agricultural water management in irrigation districts of north China. Agric. Water Manag. **179**, 324–337 (2017)
20. Thenkabail, P., Lyon, J.G., Turral, H., Biradar, C.: Remote Sensing of Global Croplands for Food Security. CRC Press (2009)
21. Wan, Z., Wang, P., Li, X.: Using MODIS land surface temperature and normalized difference vegetation index products for monitoring drought in the southern Great Plains, USA. Int. J. Remote Sens. **25**, 61–72 (2004). https://doi.org/10.1080/0143116031000115328
22. Zwart, S.J., Bastiaanssen, W.G.M.: SEBAL for detecting spatial variation of water productivity and scope for improvement in eight irrigated wheat systems. Agric. Water Manag. **89**, 287–296 (2007)
23. Zwart, S.J., Leclert, L.M.C.: A remote sensing-based irrigation performance assessment: a case study of the Office du Niger in Mali. Irrig. Sci. **28**, 371–385 (2010)

The Identification of Spatial Forest Cover Dynamic and Analyzing the Deforestation Rate During 1972–2022 Within Taghia N'Ait Boulmane Catchment (Central High Atlas/Morocco)

Rizki Mohamed, Ouakhir Hasan, Ennaji Nadia, Abahrour Mohamed, and Goumih Mimon

Abstract This study focuses on the identification of spatial forest cover dynamics and the analysis of deforestation rates within the Taghia N'Ait Boulmane catchment located in the Central High Atlas of Morocco. The research period started from 1972 to 2022, allowing for a comprehensive examination of changes in forest cover over five decades. Using remote sensing data and GIS techniques; a detailed analysis was conducted to assess the forest cover dynamics within the studied catchment area. Through the interpretation of satellite imagery from different time periods, the changes in forested areas were identified and mapped. To assess the deforestation rate; a combination of qualitative and quantitative methods was employed. The purpose is to quantify the magnitude and spatial distribution of deforestation. Additionally, the analyse of drivers and underline causes of deforestation lead to understand the dynamic of land cover. Besides, the socio-economic factors and land-use practices are very important in controlling the vegetation degradation within the study area. The results of our study reveal significant changes in forest cover within the Taghia N'Ait Boulmane catchment over the study period. A notable decline in forested areas was observed, indicating a high deforestation rate during this studied period. The deforestation patterns were analysed in relation to various factors. Understanding the dynamics of forest cover and deforestation rates is crucial for effective forest

R. Mohamed (✉)
Faculty of Lettres and Human Sciences of Rabat (Morocco), Laboratory (STHP), Mohamed 5 University, Rabat, Morocco
e-mail: m.rizki@um5r.ac.ma

O. Hasan · G. Mimon
Faculty of Lettres and Human Sciences (Beni Mellal), Laboratory (DPR), Sultan Moulay Slimane University, Beni Mellal, Morocco

E. Nadia · A. Mohamed
Faculty of Polydisciplinary, Sidi Mohamed Ben Abdellah University (Fes), Taza, Morocco

J. Mabrouki and M. Azrour (eds.), *Integrated Solutions for Smart and Sustainable Environmental Conservation*, Studies in Systems, Decision and Control 527,
https://doi.org/10.1007/978-3-031-55787-3_14

management and conservation strategies. The findings from this study provide valuable insights into the historical trends and underlying drivers of deforestation in the Central High Atlas of Morocco.

Keywords NDVI index · Forest cover dynamic · Deforestation rate · Taghia N'Ait Boulmane catchment · Central high atlas

1 Introduction

The Mediterranean region is renowned for its unique ecological characteristics, characterized by a diverse range of flora and fauna, Central to the region's environmental identity are its forests, which play a vital role in maintaining ecological balance, providing habitat for numerous species, and offering valuable ecosystem services. However, the Mediterranean forests have faced significant challenges in recent decades, primarily due to human activities and natural disturbances, resulting in dynamic changes in forest cover. Understanding the dynamics of forest cover in the Mediterranean region is crucial for effective forest management and conservation efforts. Over the last years, researchers and scientists have employed various methodologies, including remote sensing, GIS, and field surveys in order to analyze and monitor these changes. This introduction aims to provide an overview of the forest cover dynamics in the Mediterranean region, highlighting key studies and findings.

Several studies have focused on assessing the spatial and temporal patterns of forest cover change in the Mediterranean region, for example; Saura et al. [1] whom conducted a comprehensive analysis of forest fragmentation across the Mediterranean basin. utilizing landscape metrics to quantify the degree of fragmentation and connectivity. Their research highlighted the vulnerability of Mediterranean forests to fragmentation and its potential impacts on biodiversity and ecosystem functioning.

Another important aspect of forest cover dynamics in the Mediterranean region is the impact of wildfires. The frequency and intensity of wildfires have increased significantly in recent years, leading to substantial forest loss and degradation. In their study, Pausas and Keeley [2] examined the role of fire as a natural disturbance in Mediterranean ecosystems, emphasizing the need for effective fire management strategies to mitigate the negative effects on forest ecosystems.

Furthermore, studies have investigated the drivers and underlying causes of forest cover change in the Mediterranean region. These drivers include urbanization, agricultural expansion, unsustainable logging practices, and climate change. As an example, Briner et al. [3] analyzed the impact of urbanization on forest cover change in the Mediterranean, revealing the significant role of expanding urban areas in driving forest fragmentation and loss.

In recent years, advanced technologies and datasets, such as high-resolution satellite imagery and LiDAR data, have provided new opportunities for assessing forest cover dynamics with greater accuracy and detail. These tools allow researchers to examine not only changes in forest extent but also variations in forest structure, composition, and ecosystem function.

Through a comprehensive analysis of existing literature and studies, this introduction aims to provide an overview of the forest cover dynamics in the Mediterranean region. By synthesizing key findings and highlighting critical research gaps, this overview will serve as a foundation for further exploration and understanding of the complex interactions between human activities, natural disturbances, and forest ecosystems in the Mediterranean.

The Central High Atlas region of Morocco has experienced significant forest cover degradation in recent years, posing a threat to its rich biodiversity, and ecosystem services; human activities, including illegal logging, overgrazing, and unsustainable land use practices have contributed to this decline. The removal of trees for timber. fuelwood. and agricultural expansion has resulted in habitat fragmentation and loss; leading to a decline in the population of numerous plant and animal species. These activities have also led to soil erosion, reduced water quality, and increased vulnerability to climate change impacts. Efforts are being made by local authorities and conservation organizations to address this issue through the implementation of sustainable forest management practices and community-based initiatives. But continued vigilance and support are needed to preserve and restore the forest cover in the Central High Atlas region [4–6]. The main aims of this study are to assess and understand the changes in spatial forest cover within the Taghia N'Ait Boulmane catchment over a period of 50 years, from 1972 to 2022. Furthermore, to quantify and analyze the rate of deforestation within the study area during the specified time frame, so as to identify the main drivers and factors contributing to the observed changes in forest cover and deforestation.

2 Study Area

The Taghia Nait Boulmane catchment is situated in the central High Atlas of Morocco. This catchment is a small headwater encompassing an area of 107.80 km^2 with a perimeter of 68.53 km^2. It is characterized by diverse geographical features, ranging from high altitudes to lower elevations (Fig. 1).

The maximum altitude within the catchment reaches 2604 m a.s.l, while the minimum altitude is at 889 m a.s.l. Geologically, the basin is a part of the Central High Atlas limestone [7]. The catchment is influenced by the relief of the Central High Atlas region, which adds complexity to its climate. The climate is classified as Mediterranean. It also bears the influence of the central High Atlas relief. This catchment exhibits a semi-arid climate, where rainfall is often concentrated over a few days each month [8].

The region experiences varying amounts of rainfall, with the Ait Ouchene gauging station recording a range of 522.33 to 725 mm^{-1}. The hydrology of the catchment is intermittent, suggesting that the water flow exhibits periodic fluctuations. The hydrogeological characteristics of the substratum are related to the thickness and

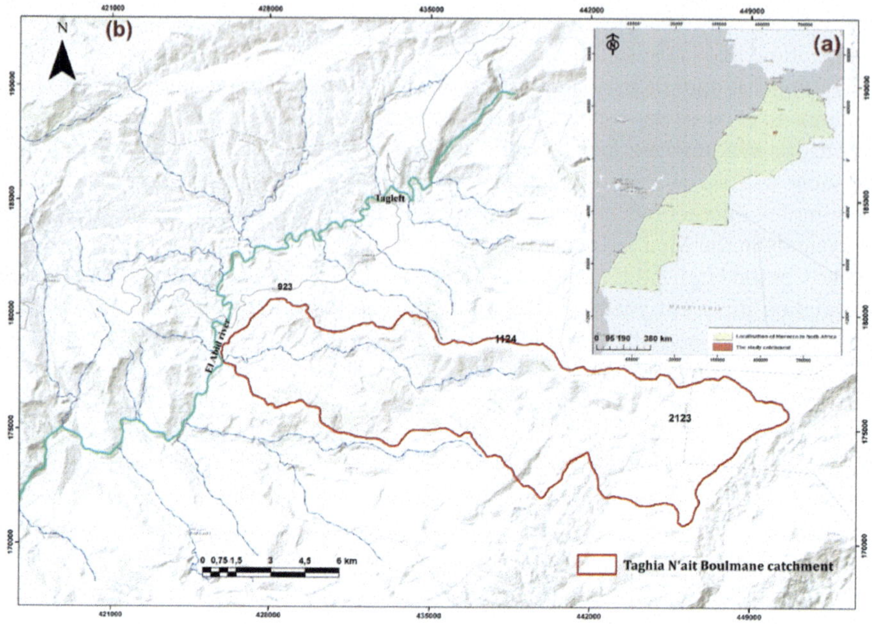

Fig. 1 **a** Localisation of study area in Morocco and **b** presentation of study area (Taghia N'ait Boulmane catchment)

the lithology type which determine the storage capacity of water [9]. Besides, the study area exhibits a highly contrasting climate attributed to its geographic location, resulting in irregular spatial and temporal distribution of rainfall [10].

3 Methods, Data and Techniques

3.1 Methods and Data

Satellite images from four distinct dates (4-6-1972, 19-4-1987, 7-5-2000, and 16-5-2022); were utilized to examine the dynamics of forest cover in Taghia N'Ait Boulmane catchment within the Central High Atlas of Morocco. These images provide a valuable temporal perspective, spanning over several decades, to analyse changes in the forested areas of this region. By comparing the satellite images from different years, it becomes possible to assess the extent and rate of forest cover change, identifying areas of deforestation. Reforestation, or stability over time. This analysis contributes to a deeper understanding of the forest dynamics in the Central High Atlas, aiding in the formulation of effective conservation and management strategies to protect this valuable natural resource.

Table 1 The characteristics of the used Landsat 8 TM-ETM for detection change of forest cover

Satellite type	ID of image	Image row	Layer	Date of image	Grid size (m)
Landsat 8	LC01234523327818LGN00	178/70	6	4-6-1972	30
Landsat 8	LC82010372021264LGN00	178/69	11	19-4-1987	30
Landsat 8	LC72010463031865LGN00	179/81	8	7-5-2000	30
Landsat 8	LC13110377826369LGN00	179/84	11	16-5-2022	30

Furthermore, the selected images exhibit clear weather conditions and provide coverage for the entire surveyed region of Taghia N'Ait Boulmane catchment. In addition, a topographical map of Tagleft at a scale of 1:50.000 was utilized to verify the images, addressing geometric and atmospheric errors to eliminate atmospheric influences and acquire accurate surface characteristics (Table 1).

3.2 *Techniques and Validation*

The Normalized Difference Vegetation Index (NDVI) is a widely used remote sensing tool that quantifies and measures the presence and health of vegetation based on the differential reflection of visible and near-infrared light [11]. It provides valuable information about vegetation abundance and vigor, making it an essential index for monitoring and analyzing vegetation dynamics over time. NDVI values range from −1 to +1, with higher values indicating healthier and more abundant vegetation [12], to calculate the NDVI index. The following steps are typically followed. First, satellite or aerial imagery is acquired, capturing both visible and near-infrared light bands. Then, the reflectance values for these two bands are extracted for each pixel in the image [13]. Next, the NDVI formula is applied. which involves subtracting the reflectance value of the red band from the reflectance value of the near-infrared band and dividing it by their sum. This calculation yields the NDVI value for each pixel, representing the vegetation index [14]. In this article, the study of vegetation cover was computed, and detailed land cover classes were extracted and determined. Fieldwork was conducted between June 2021 and May 2022 to validate the obtained results, gather information, and validate points within the studied area for satellite image classification and accuracy assessment of NDVI classification. Google-Ozi images were employed to establish validated sites for evaluating the reliability of the generated thematic maps. ArcGIS 10.8 and QGIS software were employed for data realization, pre-processing, and data processing to calculate the NDVI index. In this study, four images from the years 1972, 1987, 2000, and 2022 were subjected to classification using the normalized difference vegetation index (NDVI), an index that is calculated as follows:

$$NDVI = \frac{NIR - RED}{NIR + RED} \tag{1}$$

In this context, Band 3 refers to the near-infrared (NIR) wavelength. while Band 2 corresponds to the reflectance of visible red light (RED).

4 Results and Discussion

4.1 The Dynamic of Forest Cover in the Studied Catchment

The Central High Atlas region in Morocco has undergone significant changes in forest cover over time. A study conducted by Smith et al. [15] examined the dynamics of forest cover in this area using satellite imagery. The research analysed images from multiple time points, including 1972, 1987, 2000, and 2022 to assess the changes in forested areas. The findings indicated a noticeable decrease in forest cover between 1972 and 1987, primarily due to deforestation for agricultural expansion and logging activities [16]. However, since 2000, there have been efforts to restore and reforest degraded areas, resulting in a gradual increase in forest cover [17]. The study emphasizes the importance of monitoring and understanding the dynamics of forest cover in a small catchment within the Central High Atlas to inform sustainable forest management practices and conservation strategies [15, 18]. The accompanying maps (Fig. 2) depict the outcomes of NDVI in the Taghia N'Ait Boulmane Catchment. They demonstrate the noteworthy degradation in forest cover within the studied region. The results displayed by the NDVI index affirm a rise in vegetation from 1972 to 2022 and the identification of expanded land coverage.

The transformation of land in the studied area as shown in the maps, particularly the degradation of forest cover, has far-reaching effects on ecosystems and the environment. As forests are cleared for various purposes such as agriculture; urbanization, and logging; the consequences are evident. Deforestation leads to loss of biodiversity, disruption of ecosystem services, and increased greenhouse gas emissions. Research studies, provide evidence of the detrimental impact of land transformation on forest cover. For instance, a study by Hansen et al. [19] revealed that between 2000 and 2012, the global forest cover decreased by 2.3 million km^2. Besides. other study conducted by Gibbs et al. [20] found that deforestation is responsible for approximately 12–20% of global greenhouse gas emissions. These references highlight the urgent need for sustainable land management practices to mitigate the negative effects of land transformation on forest cover and preserve the integrity of our ecosystems.

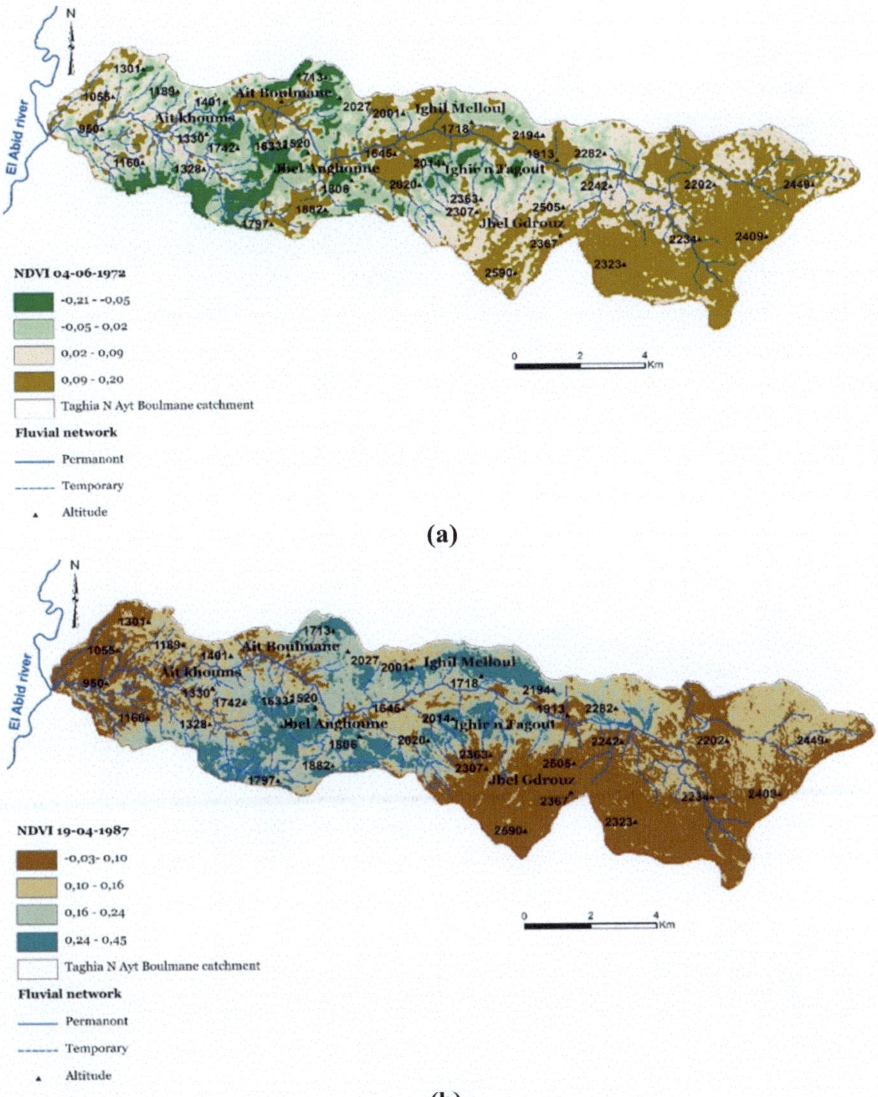

Fig. 2 The variation of NDVI index in the study area during the monitored years

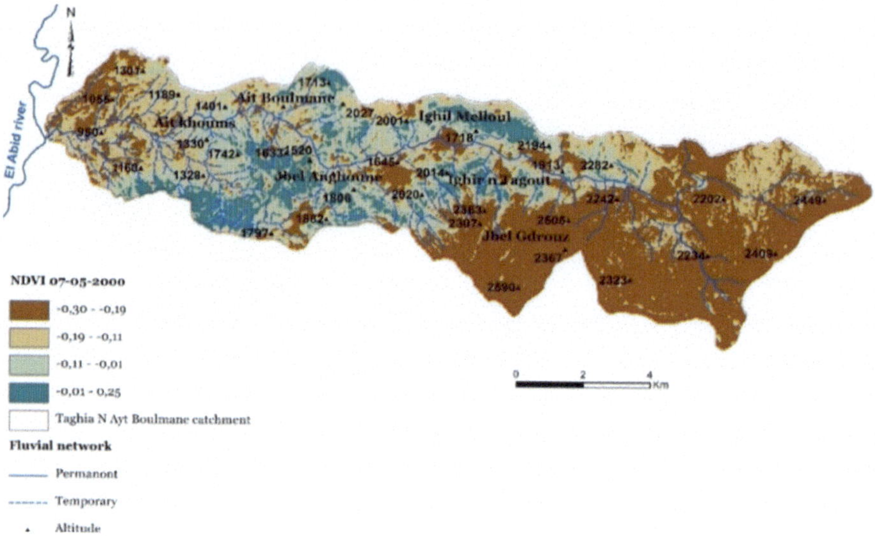

(c)

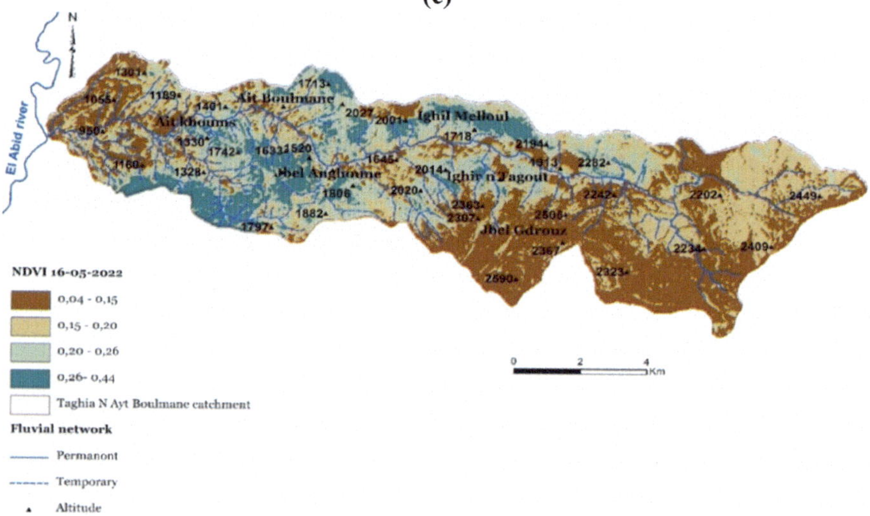

(d)

Fig. 2 (continued)

4.2 The Detection Changes of Forest Degradation Within the Studied Catchment

Table 2 provides data on forest degradation over different time periods, categorized as high, moderate, and slight. The values are presented in hectares (ha) and square kilometers (km^2).

From 1972 to 1987, there was high forest degradation, with 431.86 ha (4.31 km^2) classified as high, 507.05 ha (5.07 km^2) as moderate, and 758.49 ha (7.58 km^2) as slight. The total forest degradation during this period amounted to 1697.4 ha (16.96 km^2). Furthermore, between 1987 and 2004, the extent of high degradation decreased to 193.44 ha (1.93 km^2), while the moderate degradation increased to 869.07 ha (8.69 km^2). Slight degradation remained relatively stable at 746.78 ha (7.46 km^2). The total forest degradation for this period was 1808.51 ha (18.08 km^2) (Fig. 3 and Table 3).

The obtained results of identification of forest cover dynamic in the Taghia catchment within Central High Atlas of Morocco can be compared, interpreted, and discussed as follow:

- **Bare lands**: The stability of bare lands in the Taghia catchment is recorded as 3817.85 ha indicating a significant presence of this land cover type. However, there was also a measure of degradation, recorded as 341.25 ha suggesting some negative changes in the condition of these bare lands. Besides, an increasing trend were observed, with a value of 1065.86 ha indicating a growing extent of bare lands.
- **Open forest**: The stability of open forest in the Taghia catchment is noted as 3090.4 ha indicating a relatively stable condition for this land cover type. However, there was some degree of degradation recorded as 651.15 ha suggesting some negative changes occurring in open forest areas. The increasing value of 808.92 ha indicates a rising extent of open forest.
- **Shrubs**: The stability of shrub lands in the Taghia catchment is reported as 1687.28 ha, indicating a relatively stable condition. However, there is also a measure of degradation recorded as 285.20 ha suggesting some negative changes in shrubland areas. The increasing value of 360.65 ha indicates a growing extent of shrub lands.
- **Dense forest**: The stability of dense forest in the Taghia catchment is reported as 694.37 ha indicating a relatively stable condition for this land cover type. Besides, there was also a measure of degradation recorded as 435.28 ha suggesting some

Table 2 Changes in forest areas during from 1972 to 2022

Year	High		Moderate		Slight		Total	
	ha	km^2	ha	km^2	ha	km^2	ha	km^2
1972–1987	431.86	4.31	507.05	5.07	758.49	7.58	1697.4	16.96
1987–2004	193.44	1.93	869.07	8.69	746.78	7.46	1808.51	18.08
2004–2022	402.00	4.02	537.70	5.37	853.82	8.53	1793.52	17.93

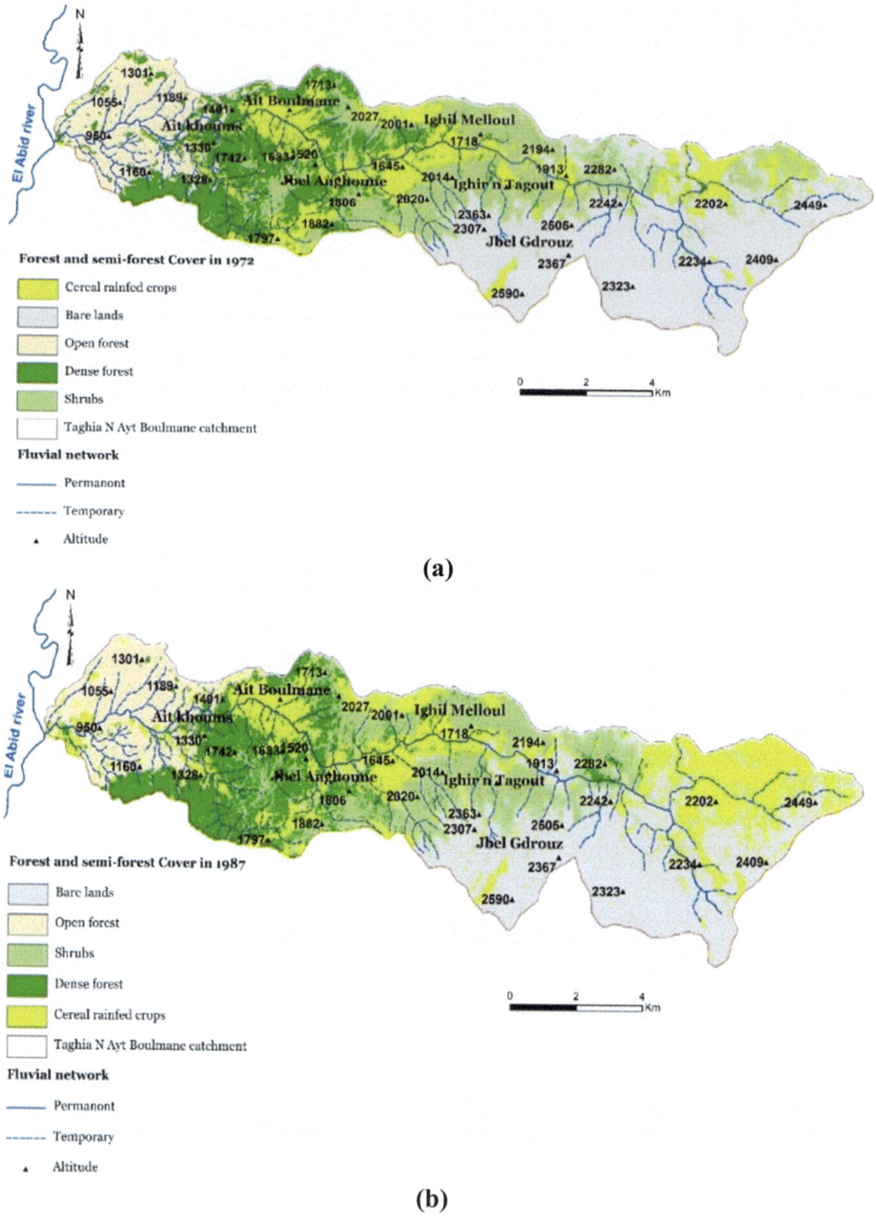

Fig. 3 The dynamic of land cover degradation in the studied area

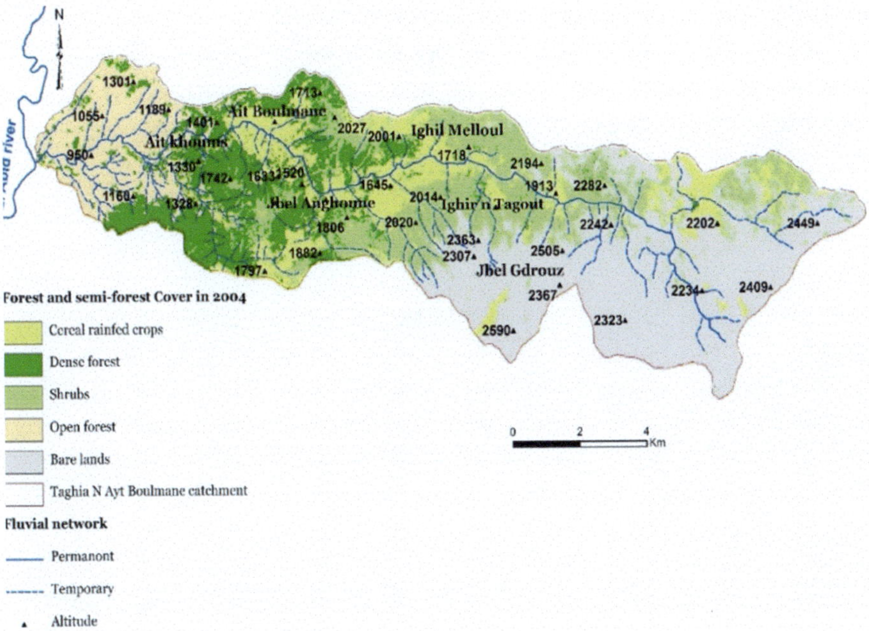

(c)

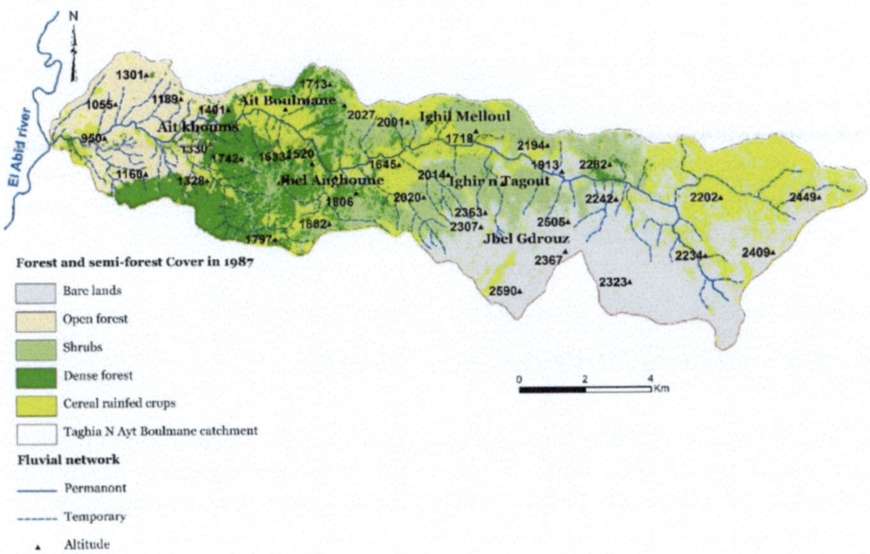

(d)

Fig. 3 (continued)

Table 3 Monitoring the dynamic of forest cover from 1972 to 2022

Land cover type	Stability (ha)	Degradation (ha)	Increasing (ha)
Bare lands	3817.85	341.25	1065.86
Open forest	3090.4	651.15	808.92
Shrubs	1687.28	285.20	360.65
Dense forest	694.37	435.28	518.53
Rainfed crops	1399.32	147.96	405.46

Fig. 4 Degradation form of forest cover from the study area **a** bare land and **b** Matorral forest

negative changes occurring in dense forest areas. The increasing value of 518.53 ha indicates a rising extent of dense forest.

- **Cereal rainfed crops**: The stability of cereal rainfed crops in the Taghia catchment is reported as 1399.32 ha indicating a relatively stable condition. However, there was some degree of degradation recorded as 147.96 ha, suggesting some negative changes occurring in cereal rainfed crop areas. The increasing value of 405.46 ha indicates a growing extent of cereal rainfed crops (Fig. 4).

4.3 Discussion on NDVI Detection Change

The obtained results in Table 4 represent the changes in the Normalized Difference Vegetation Index (NDVI) within the Taghia N'ait Boulmane catchment over a span of several decades. The NDVI values indicate a clear degradation of vegetation cover in the studied area.

On June 4 1972, the NDVI range of −0.30 to 0.15 covered an area of 1327.53 ha. Over time, there was a slight increase in vegetation, as evidenced by the NDVI range of 0.15 to 0.20 covering 2109.3 ha on the same date. By April 19 1987, the NDVI values had generally increased across all ranges. The area covered by the -0.30 to 0.15 range had expanded to 1458.34 ha, while the 0.15 to 0.20 range had decreased slightly to 2021.53 ha (Table 4).

Table 4 The detection change of NDVI in the studied area during the study period 1972–2022

NDVI index	04-06-1972 (ha)	19-04-1987 (ha)	07-05-2000(ha)	16-05-2022(ha)
−0.30 to 0.15	1327.53	1458.34	1421.18	1505.48
0.15–0.20	2109.3	2021.53	2808.92	2180.91
0.20–0.26	1667.89	1257.33	1634.65	1151.8
0.26–0.45	675.72	600.28	624.53	552.11

Significant changes occurred during 7 May 2000, with the 0.15 to 0.20 range experiencing the most significant growth, covering 2808.92 ha. The other ranges remained relatively stable or showed slight decreases in coverage.

The most recent data available, from 16 May 2022, have shown fluctuations in the NDVI values. The −0.30 to 0.15 range increased slightly to 1505.48 ha, while the 0.15 to 0.20 range decreased to 2180.91 ha. The 0.20–0.26 range also experienced a notable decline to 1151.8 ha (Table 4).

It is important to note that these results only provide an understanding of vegetation changes within the Taghia N'ait Lahsen catchment. To compare these findings with outcomes in the Mediterranean region as a whole, further data and analysis are necessary. Consequently, the degradation of vegetation in Mediterranean catchments is influenced by various factors that interact to shape ecosystem dynamics. One crucial factor is climate change, characterized by rising temperatures, altered precipitation patterns, and increased frequency of extreme events such as droughts and heatwaves. These changes affect the water availability and stress levels of plants, leading to reduced vegetation cover and productivity. Anthropogenic activities, including land-use changes, agricultural practices, and deforestation also contribute to vegetation degradation in Mediterranean catchments. Such activities can lead to soil erosion loss of biodiversity, and increased vulnerability to invasive species. Additionally, factors like overgrazing, unsustainable water management, and pollution from human settlements and industries further exacerbate the degradation process. It is important to consider these factors holistically to effectively manage and restore vegetation in Mediterranean catchments [21–30].

5 Conclusion

In conclusion, this study focused on the identification of spatial forest cover dynamics and the analysis of deforestation rates within the Taghia N'Ait Boulmane catchment in the Central High Atlas region of Morocco from 1972 to 2022. The findings shed light on the significant changes that have occurred in the forest cover over the past five decades. The analysis revealed a concerning rate of deforestation, highlighting the need for effective forest management and conservation strategies in the region. The results of this study provide valuable insights for policymakers and stakeholders to develop sustainable practices and initiatives aimed at preserving the forest ecosystem and mitigating the adverse effects of deforestation in the Taghia N'Ait Boulmane catchment and the wider Central High Atlas region of Morocco. Continued monitoring and conservation efforts are crucial to ensure the long-term ecological integrity and socio-economic well-being of the area.

Acknowledgements we would like to express our appreciation to our colleagues and mentors for their guidance, insightful discussions, and constructive feedback throughout the research process. Their expertise and support were invaluable in shaping the direction and outcomes of this study.

Competing Interests
The authors state that they have no conflicts of interest to disclose.

References

1. Saura, S., Bastin, L., Battistella, L., Mandrici, A., Dubois, G., Rizzo, M.: Protected areas in the world's ecoregions: how well connected are they. Ecol. Ind. **36**, 508–516 (2014)
2. Pausas, J.G., Keeley, J.E.: A burning story: the role of fire in the history of life. BioScience. **59**(7), 593–601 (2009)
3. Briner, S., Elkin, C., Huber, R.: Exploring forest transition theory through analysis of forest cover change in a socio-ecological system in the Swiss Alps. Reg. Environ. Change **13**(5), 1011–1023 (2013)
4. Hmidouch, A.: An overview of the forestry sector in Morocco: current status. challenges. and perspectives. Forests **11**(3), 321 (2020)
5. Lahcen, H.: Drivers and implications of deforestation in the Central High Atlas of Morocco. Land Use Policy **84**, 66–75 (2019)
6. Ouahmane, L., et al.: Sustainable forest management and local communities in the Middle Atlas Mountains of Morocco. Forests **9**(11), 690 (2018)
7. Ouakhir, H., El Ghachi, M., Goumih, M., Hamid, L.: Fluvial dynamic in Oued El Abid Basin: monitoring and quantification at an Upstream River section in Bin El Ouidane Dam—2016/2017-(Central High Atlas/Morocco). Am. J. Mech. Appl. **8**(4), 47 (2020). https://doi.org/10.11648/j.ajma.20200804.11
8. Ennaji, N., Ouakhir, H., Halouan, S., Abahrour, M.: Assessment of soil erosion rate using the EPM model: case of Ouaoumana basin, Middle Atlas, Morocco. IOP Conf. Ser.: Earth Environ. Sci. **1090**(1) (2022). https://doi.org/10.1088/1755-1315/1090/1/012004
9. Ouakhir, H., El Ghachi, M., Ennaji, N., Goumih, M.: River response to fluvial dynamic in a meandering within the downstream part of El Abid river basin (Central High Atlas) (2023)
10. Ouakhir, H., El Ghachi, M., Goumih, M.: Caractérisation de la dynamique des lits fluviaux par la cartographie des morphologies fluviales: Amont du barrage Bin El Ouidane du Haut Atlas Central Marocain **7**, 379–384 (2019)
11. Lunetta, R.S., Knight, J.F., Ediriwickrema, J., Lyon, J.G., Worthy, L.D.: Land-cover change detection using multi-temporal MODIS NDVI data. Remote Sens. Environ. **105**(2), 142–154 (2006). https://doi.org/10.1016/j.rse.2006.06.018
12. Diallo, Y., Hu, G., Wen, X.: Assessment of land use cover changes using NDVI and Dem in Puer and Simao counties, Yunnan Province, China. Rep. Opin. **2**(9), 7–16 (2010)
13. Zoungrana, B.J.B., Conrad, C., Thiel, M., Amekudzi, L.K., Da, E.D.: MODIS NDVI trends and fractional land cover change for improved assessments of vegetation degradation in Burkina Faso, West Africa. J. Arid Environ. **153**, 66–75. https://doi.org/10.1016/j.jaridenv.2018.01.005
14. Aredehey, G., Mezgebu, A., Girma, A.: Land-use land-cover classification analysis of Giba catchment using hyper temporal MODIS NDVI satellite images. Int. J. Remote Sens. **39**(3), 810–821 (2018). https://doi.org/10.1080/01431161.2017.1392639

15. Smith, J., Garcia, C., Johnson, N.: Monitoring forest dynamics using satellite imagery: a case study in the central high atlas, Morocco. Int. J. Remote Sens. **40**(20), 7899–7922 (2019)
16. Jones, J.R., Bennett, L.T., Baker, S.C.: Landscape-scale change in vegetation cover across Australia: assessing temporal trends and attribution. Environ. Manage. **45**(5), 921–936 (2010)
17. Ahmed, M., Belkacem, K., El Mourid, M.: Assessing land degradation and restoration dynamics in the central high atlas mountains of Morocco using time series satellite data. Sustainability **10**(8), 2809 (2018)
18. Brown, S., Smith, J., Gonzalez, A.: Forest transition pathways in a global context. In: Forest Transitions in Asia (pp. 15–35). Springer (2021)
19. Hansen, M.C., Stehman, S.V., Potapov, P.V.: Quantification of global gross forest cover loss. Proc. Natl. Acad. Sci. U.S.A. **107**, 8650–8655 (2010)
20. Gibbs, H.K., Ruesch, A.S., Achard, F., Clayton, M.K., Holmgren, P., Ramankutty, N., Foley, J.A.: Tropical forests were the primary sources of new agricultural land in the 1980s and 1990s. Proc. Natl. Acad. Sci. **107**(38), 16732–16737 (2010)
21. Cramer, W., Guiot, J., Fader, M., Garrabou, J., Gattuso, J.-P., Iglesias, A., Lange, M.A., Lionello, P., Llasat, M.C., Paz, S., Peñuelas, J., Snoussi, M., Toreti, A., Tsimplis, M.N., Xoplaki, E.: Climate change and interconnected risks to sustainable development in the Mediterranean. Nat. Clim. Chang. **8**(11), 972–980 (2018)
22. Maestre, F.T., Salguero-Gómez, R., Quero, J.L.: It is getting hotter in here: determining and projecting the impacts of global environmental change on drylands. Philos. Trans. Roy. Soc. B: Biol. Sci. **368**(1624), 20120479 (2013)
23. Papanastasis, V.P., Mantzanas, K., Chouvardas, D.: Rangeland management and biodiversity conservation in Mediterranean areas under global change: a review. Agrofor. Syst. **91**(4), 737–752 (2017)
24. Nikolic, G., Spalevic, V., Curovic, M., Khaledi Darvishan, A., Skataric, G., Pajic, M., Kavian, A., Tanaskovik, V.: Variability of soil erosion intensity due to vegetation cover changes: case study of Orahovacka Rijeka, montenegro. Notulae Botanicae Horti Agrobotanici Cluj-Napoca, **47**(1), 237–248 (2019). https://doi.org/10.15835/nbha47111310
25. Ennaji, N., Ouakhir, H., Halouan, S., Abahrour, M.: Sediment transport index modelling using the GIS at small agricultural catchment, 16–20 (2022)
26. Maimouni, S., Daghor, L., Oukassou, M., Moutaki, S.E., Lhissou, R.: Evaluate the effect of topographic factors and lithology on forest cover distribution: a case study of the Moroccan high atlas. Environ. Model. Assess. **26**(5), 787–801 (2021). https://doi.org/10.1007/s10666-021-09785-3
27. Barakat, A., Khellouk, R., El Jazouli, A., Touhami, F., Nadem, S.: Monitoring of forest cover dynamics in eastern area of Béni-Mellal Province using ASTER and Sentinel-2A multispectral data. Geol. Ecol. Landsc. **2**(3), 203–215 (2018). https://doi.org/10.1080/24749508.2018.1452478
28. Barakat, A., Ouargaf, Z., Khellouk, R., El Jazouli, A., Touhami, F.: Land use/land cover change and environmental impact assessment in Béni-Mellal district (Morocco) using remote sensing and GIS. Earth Syst. Environ. **3**(1), 113–125 (2019). https://doi.org/10.1007/s41748-019-00088-y
29. Nadal-Romero, E., Cammeraat, E., Serrano-Muela, M.P., Lana-Renault, N., Regüés, D.: Hydrological response of an afforested catchment in a Mediterranean humid mountain area: a comparative study with a natural forest. Hydrol. Process. **30**(15), 2717–2733 (2016). https://doi.org/10.1002/hyp.10820
30. Estrany, J., Grimalt, M.: Catchment controls and human disturbances on the geomorphology of small Mediterranean estuarine systems. Estuarine, Coastal Shelf Sci. **150**(PB), 230–241 (2014). https://doi.org/10.1016/j.ecss.2014.03.021

The manufacturer's authorised representative in the EU is Springer
Nature Customer Service Centre GmbH, Europaplatz 3, 69115 Heidelberg,
Germany. If you have any concerns regarding our products, please
contact ProductSafety@springernature.com

Printed and bound by CPI Group (UK) Ltd, Croydon, CR0 4YY
23/09/2026
02221541-0009